# YOUR COUNTRY COTTAGE

## A Guide to Purchase and Restoration

# YOUR COUNTRY COTTAGE

## A Guide to Purchase and Restoration

R. C. EDMUNDS

DAVID & CHARLES
NEWTON ABBOT

7153 4794 2



Set in twelve on thirteen point Baskerville
and printed in Great Britain
by Latimer Trend & Company Limited Plymouth
for David & Charles (Publishers) Limited
South Devon House Newton Abbot Devon

The old farms and cottages scattered up and down the land possess an essential requirement of successful building in that they appear to be as much part and parcel of the landscape as the hedges and trees. And yet it is unlikely that the beauty of these buildings was apparent to their builders. Their beauty came as the spontaneous products of the hands of their constructors, who were in the stage of culture in which technical ability produces works of art naturally and unconsciously, unlike the technically skilled workers of present day civilisation, who can only design 'works of art' after much training. Our smaller buildings became ugly when technique had become highly skilled and the workers had mastered their materials completely.

C. F. Innocent
*Development of English*
*Building Construction*
(1916)

The homely old English cottages were models of architecture in their way.

William Morris

In many of the villages and in the countryside of England old cottages are becoming derelict, and being demolished or falling down. Many of these are buildings of real merit and beauty, and are part of the historical background of the country; the destruction of others makes unseemly interruptions in the village scene.

Representatives of the Council for the Preservation of Rural England, the National Association of Parish Councils and the Pilgrim Trust have decided, after discussion with the Ministry of Housing and Local Government, to carry out a survey of the causes of this problem and possible remedies.

Extract from press notice
issued by the Secretary to the
National Association of Parish
Councils, April 1967.

# Contents

# Illustrations

PLATES

IN TEXT

# Introduction

This book is for those who would love to buy an old cottage in the country and 'do it up', but feel they are too inexperienced to risk it. They fear the probable cost of having to trust strange builders with the job of restoration. They are daunted by the well meaning caution of most surveyors, who prophesy dry rot, rising damp, and endless expense. A sort of mystique has arisen around the restoration of old property; a hands-off, this-is-for-the-expert-only, attitude that deters the home buyer who would love to have a go. So he plays safe and buys a new bungalow. And somewhere an old cottage that would have become a hobby as well as a home crumbles back into the soil.

The thing to remember about these 'risky' old houses is that they have been standing for 200–300 years. Most of them were built by hard-headed practical men who chose the most advantageous site at a time when land was plentiful. They were built personally—often by craftsmen. There was no mass production, no dubious factory output of unseasoned wood; every beam was hand-adzed, the larger beams, or summer beams, in many cases having been left in a bog for up to four years to 'cure'. These big load-carrying beams are usually heart of oak, meaning that they have been cut an inch and a half inside the sap-line to ensure they were resistant to rot or insects. Nothing much can happen to solid cob or stone walls, providing they have not been exposed to years of constant wet soaking in from the top. As for rising damp, this is curable, make no mistake.

The first old property my wife an I bought was a derelict inn—for £170, which gives some idea of the state it was in. At that time we had nothing but enthusiasm to go on, and an idea of what we wanted backed by some knowledge of period architecture. It cost us a fair bit to gain the practical experience we have now, but it was worth it. We called in a builder to do the necessary structural work, and did the remainder ourselves; and the 'remainder' would have been very expensive otherwise, as it was mostly labour—scraping beams, mending windows, exposing woodwork, decorating, etc. It was hard work, but immensely enjoyable as we were in no great hurry. We had to sell the house eventually as it became too far for the children to travel to and from school. Recently, however, we heard that—without much extra work being done on it apart from general maintenance—it sold for nearly two and a half times what it cost us to buy and restore it. Our first reaction, believe it or not, was pride that we must have done a pretty good job. Since then we have had real satisfaction and pleasure in restoring many otherwise doomed old houses, including an historic little courthouse in Queensland, Australia.

Most of what we have learned has been in the hard school of experience and from the sort of people who built these houses originally—village craftsmen, old stone-masons, thatchers, local carpenters and joiners. This experience, added to a continuing study of buildings of the period, has enabled us to know what to look for under the present-day exterior, and how to restore it to something like its original appearance, besides giving it modern amenities.

In the past few years we have also found ourselves advising people who want to buy an old country property and restore it. Their problems, and possibly yours, are how to find the right place, how to obtain advice on reorganising the interior, and, above all, reassurance born of experience that it will turn out all right.

CHAPTER 1

# What to Look For

The first requisite in buying a derelict old cottage in the country and restoring it as a comfortable place to live in is, of course, some money. The second is love: you must be able to love that heap of old rubble and broken windows. The third is imagination: the ability to imagine what it will look like as the result of your money, your love and, fourth requisite, your hard work.

It is fairly safe to say that you can buy an old cottage and restore it to habitable condition for about what you would pay for a new bungalow. But remember that the former will have a mature garden, already fenced. If you must have a mortgage, the situation is rather more tricky but not insurmountable. For the same amount that you would have to put down on a newer property, say £1,500 to £1,750, you should be able to buy outright a cottage needing restoration; and if you are lucky, for very much less. Then by depositing the deeds with your bank it should be possible, in normal times, to obtain the amount needed to carry out the necessary work. This assumes, of course, that you can convince your bank manager that you are a good risk—but a Building Society would need that reassurance anyway. You will also be in a position to apply for a housing grant, which we will go into later.

It is, however, almost impossible to draw an exact parallel between the old and the new, as much depends on location and size. As a rule properties tend to be cheaper the farther one goes from London or other metropolitan centres, and site value is usually the big factor in buying derelict property. Quite often you do not pay for the building as such, you pay for the ground it stands on. In this you have one advantage over the new house builder. If he fancies the site which is already occupied by an old building he must pull down the old house first and dispose of the remains. This can be an expensive item.

Size is no problem. You can usually get three bedrooms—two moderate and one small—out of the average country cottage—more out of an old farmhouse—plus lounge, dining-room, kitchen, and bathroom: about the size of the average bungalow. These comparisons are mentioned as an indication of price, not choice. If you are going to embark on this project and see it through it must be because you want to live in the sort of house that has character and atmosphere belonging to perhaps centuries of habitation. You will need to be dedicated, because there are bound to be times when you will wonder if it is all worth it. Only if you are dedicated will you know that it is—a hundred times over.

Even in the West Country the price of old cottages

*Page 17* Fourteenth-century yeoman's house (Somerset) before and after restoration, (*above*) front before; (*below*) front after

*Page 18* Fourteenth-century yeoman's house (Somerset) before and after restoration, (*above*) rear before; (*below*) rear after

has skyrocketed in the last few years, but it is still possible, if you keep your eyes open, to find one at a reasonable price. Agents are not usually the best people to approach—they tend to advise owners to overprice or, alternatively, if the owners approach the agents in the first place they often have over-optimistic ideas themselves. It is, however, a good idea to take the local paper regularly as properties are sometimes advertised privately.

Personally we have found that the best way is to go out and look. If you live too far away, then it might be worth spending a holiday in the area of your choice, simply driving around the less-frequented roads and keeping a sharp lookout. The village pub, where local ears are close to the ground, is often a good source of information; we found two cottages by just this method of sniffing around. Both had demolition orders on them; both belonged to farmers.

Here it might be worth while to mention farmers in relation to old cottages, as outside the villages most of them go together. The majority of these cottages were originally built as farmworkers' habitations, but sometimes, in taking over a neighbouring property, the farmer has no use for the farmhouse that goes with it and either uses it for storage, disposes of it, or, too commonly, lets it fall down. Unfortunately, farmers generally dislike disposing of any of their property and would rather see a house fall down than sell off a small slice of land that might be in the middle of their acreage.

In order to use the cottage or farmhouse to rehouse his own workers a farmer must satisfy the local authorities that it conforms to their housing standards. Rather than spend money on doing this, he will often prefer to pull down the sturdy old cottage and build an eyesore in its place. So he lets the matter slide until the council first declares the cottage unfit for habitation and puts an 'Undertaking Order' on it; and then, as its state progressively worsens, finally extends that to a 'Demolition Order'. The latter usually carries a time limit, and on its expiry the farmer is compelled to demolish the building or face a fine.

A 'Demolition Order' is not as final nor as fearsome as it sounds. Most rural councils, despite their reputation as cold-blooded bureaucrats, are only too anxious to preserve the character of the countryside, and are very willing to co-operate with anyone ready to undertake the work and expense of restoration. Two rural councils even worked with us to the extent of advising us what condemned or 'closed' properties there were in their areas, in the hope that they could be saved in cases where the owners were unwilling or unable to do anything about them.

So, if you find an empty old cottage and it belongs to a farmer, it is the farmer you will have to woo. At first he will probably not even talk to you. We spent two weeks following one old farmer around without even getting him to discuss the matter. Then, as his demolition deadline drew nearer, we could sense him weakening. It might seem an easy proposition to knock down an old cottage, particularly one built of cob—a pounded mixture predominantly of mud and straw. But no! A demolition man we know spoke despairingly of his experiences pulling down cob structures with 3 ft thick walls: he threw a steel hawser round the house, attaching it to his tractor and hauling away, confidently expecting the poor old ruin of a house to subside into the dust; only to find that the hawser merely cut clean through the walls, which dropped an inch and settled comfortably back into position.

Our farmer was eventually cornered in one of his barns. He drew an outline of the land he was prepared to sell in the thick clover seed on the floor. We haggled a few clover seeds back and forth and that was that. He stuck faithfully to his bargain and in due course it was translated into legal documentation. The house, a fifteenth-century yeoman's dwelling, containing some magnificent cross-beams, panelling, and a rare early example of a dog-stair, was of no value to him whatsoever. But he thought he could have got planning permission for two small houses on the site, so he priced it at £700. Had we purchased it through a third party, or had it been put up for sale on the open market, we no doubt should have paid considerably more.

This is not to say it is easy to acquire an old cottage in this manner. A Somerset farmer we approached had a beautiful little sixteenth-century house sitting by itself in a cottage garden surrounded by open fields. The house was derelict and he was unconcernedly watching it fall down, but the field it cut into was his field and though there was a road passing the house, he did not want strangers living so near his land. He had had very high offers for this particular piece, about a quarter of an acre, but he refused even to consider them.

Another prosperous farmer told us a cottage he owns, which is almost derelict and fronts on a main road, is worth £12,000 to him. It lies in the centre of his huge farm and to sell even half an acre of land with it would reduce the value of his property by £10,000. So one day this charming old place, once an inn and staging post, now a farmworker's dwelling which the council are already eyeing with suspicion, will be declared unfit for habitation, the roof will fall in, and wind and weather will do the rest.

Rather different is the situation where the farms are owned by a county council's Smallholding Department. Occasionally, when the farmhouses fall into a bad state of repair, the council will build a new house for the tenant farmer. If the old house had character at all it is usually put up for auction, with a suitable area of land for a garden. So it is worth looking out for these auctions in the local paper, or contacting the local Smallholdings Department direct.

The position within villages is somewhat different. Cottages in villages, not already privately owned, as a rule belong to the estate of the local squire, whose family at one time owned the whole village and probably several others in the vicinity. Death duties and economic pressures have resulted in the breaking up of many of these little domains, and from time to time they come on the market for auction, parcelled up into farming lots, with the village cottages, and perhaps an inn, or smithy, or carpenter's shop, being sold off separately. Sometimes these cottages are in fair condition, if they have been lived in recently; sometimes they come up for sale with the tenants living in them; sometimes, like the farmers' cottages, they have been allowed to fall into complete disrepair by a disinterested landlord. And sometimes, when the whole estate is entailed, the landlord can neither sell the cottages nor afford to repair them. One village in Hampshire can boast of eighteen of these sad, empty little houses, put by a dead hand out of everyone's reach.

Some universities own old village properties and might well be persuaded to sell. In such cases they usually like to have some assurance beforehand that the purchaser will, if any alterations are made, retain the intrinsic character of the property.

It is well to advise one or two of the bigger London estate agents that you wish to be kept informed of family estate auctions. Among village or nearby property, inns are most sought after. If they have been converted to private use before the rise of the great brewery houses they will probably still belong to the estate, or to private owners; and they will be in much better shape for your purposes, provided the owners have left them alone structurally. Once the brewery houses start adding gimmicky fireplaces, vinyl floors, varnished panelling, and boxed-in beams, they will want a lot of money to let you have the privilege of taking them all out.

While you are inspecting the various lots up for sale, don't overlook the aforementioned smithy or carpenter's shop. It may be tiny, but a busy smithy in a largish village will be big enough for you to turn into an interesting little house. These buildings are nearly always one-storey, with a massive chimney and forging bench in front of it, and usually a cobbled yard outside. They are hardly high enough to get two stories, though sometimes you can make enough space by putting in dormer windows. But the ground floor, with perhaps a lean-to added, should give two bedrooms and a sitting-room and usual offices. The carpenter's shop is not as a rule such a good proposition, unless you want a really small hideaway, but they tend to be cheaper, besides being centrally located in the village.

Now a word about old mills. Nearly everyone this way inclined has romantic notions about living in an old converted mill, with the millstream babbling by.

But there is another side to the picture. Mills were built as close to the water and as low on the landscape as they could be, for obvious reasons. Most mills actually have a stream running underneath them. So to begin with you have a damp problem that is intrinsic and persistent. Apart from the damp, mills are very individual structures and very expensive to convert. There is, of course, the miller's cottage, which is usually part of the main building and can be restored fairly easily; but then you have a small cottage a fraction the size of its huge appendage.

To convert the mill itself means more or less ripping out its whole interior, as it will consist of several different floor levels containing a complex of milling processes, constructed of vast, often diagonal, beams and cross-supports. If you are willing and rich enough to do this, you can build a delightful house out of the shell. But damp may remain a continuing problem.

It is worth investigating the possibilities of almost every disused old building that comes on to the market, or that you see standing empty, if you like the look of it and its surroundings. Old chapels are good subjects, but make sure there is not a burial ground adjacent. Not for fear of being haunted, but because you cannot dig up graves or build over them until 100 years have elapsed since the last body was interred; even if you wanted to! It is possible to seek permission from a responsible relative to have the coffin removed and re-interred, but it is not an attractive prospect. We nearly bought a beautiful little Regency chapel in what looked like a wild garden—before closer investigation revealed the fallen tombstones under the weeds, the latest of which was dated 1946.

The latest offerings in redundant old buildings are

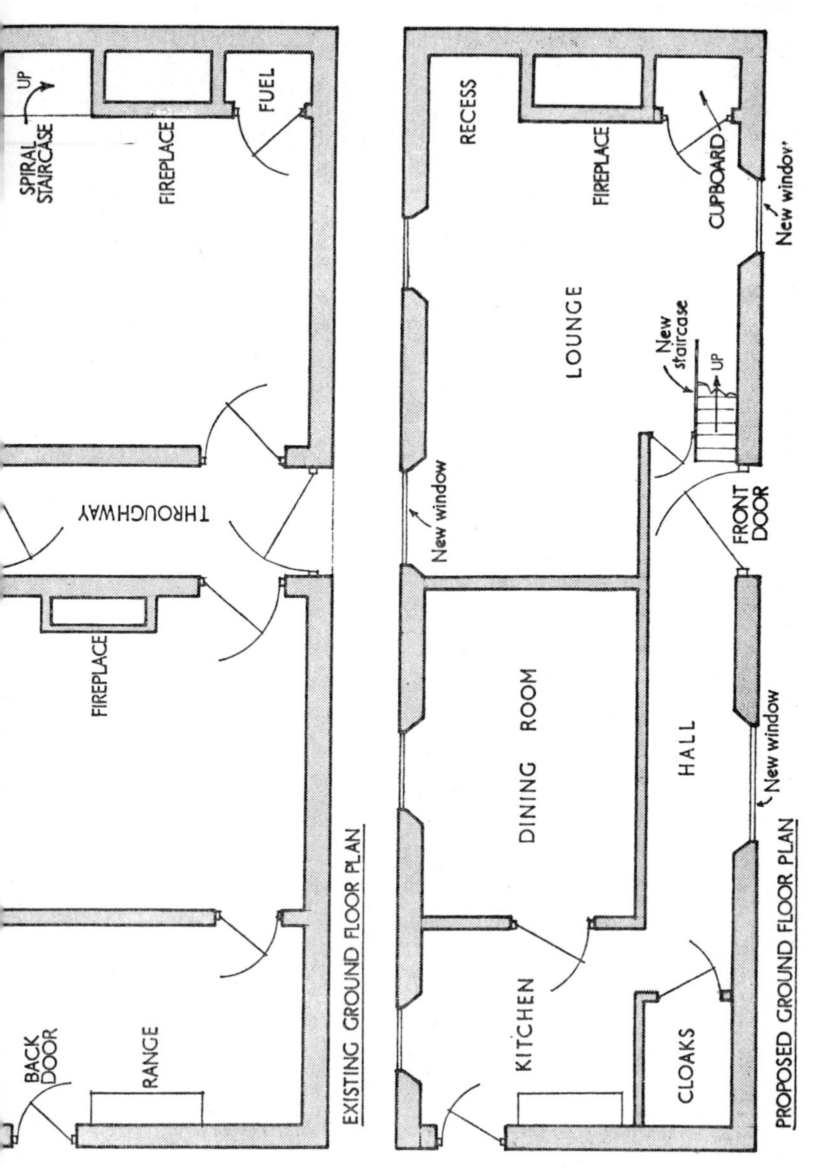


Rough plan of conversion of old farmhouse

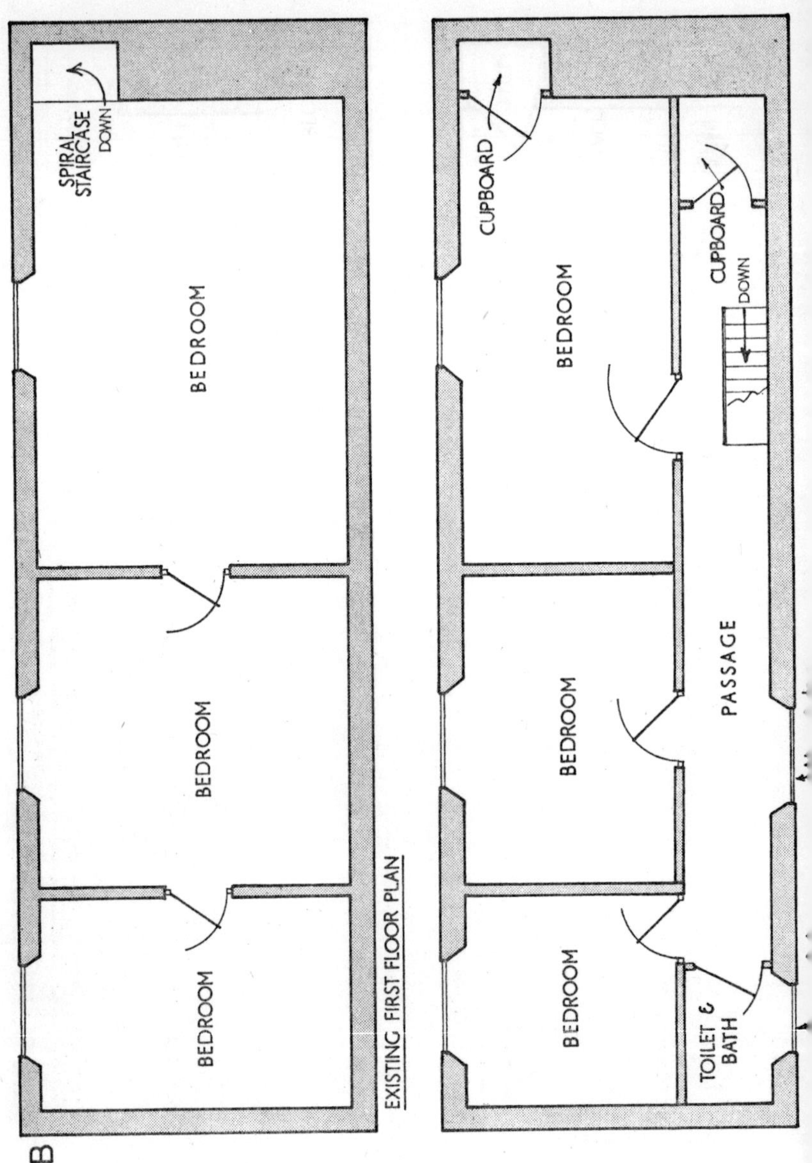


Rough plan of conversion of old farmhouse

built in Tudor times, when the intending occupants wanted an improvement on the one-room halls of their forbears, or built an extension of the original hall itself. It is interesting to note that the old farm-houses of any quality, still in use today, which were built before the eighteenth century, are most likely to have been small manor houses originally. Before farming land was parcelled into neat acreages the farmer himself probably lived in the nearest village.

The entrance to the farmhouse is usually by way of an external porch, through what was originally a solid plank door, into a flagged hallway some 5 ft wide. There will possibly be a door on either side, one leading to the parlour with fireplace, the other to a former kitchen with a fireplace of more impressive proportions. Smaller rooms may lead off from either of these, likewise the staircase. Above will be three or four bedrooms, usually with intercommunicating doors. If this sounds rather over-simplified remember that there were no professional architects until the middle of the sixteenth century, and then only the wealthy could afford their services. If a farmer wanted a house built he called in the local handyman and told him how much accommodation he needed; and the handyman built him a replica of the last house he had built, with modifications, the original plan for which was based on a simplified version of the big houses being constructed for the more wealthy gentry in the vicinity.

The cottages were almost as standard in layout as a row of working-class houses today, except that the occupant added his own personal touch to them—if he was a thatcher it had a particularly rich thatch on it, if he were a plasterer he usually ornamented his own walls with his craft; and the same with the mason and the woodcarver. If you are lucky you can still find evidence of this sort of work. Each cottage achieved its own character, which is more than can be said of present-day small buildings.

Thus we get an area blueprint for houses of comparable size built or added to during the same period. Extracts from contracts drawn up between householder and workman in the seventeenth century illustrate this:

> Great Dalby (1690): . . . a house and Parlour with Chambers over them containing two bayes of building; a lean-to adjoining to the house for a small room. A barn . . .
>
> Contract in Leicester Museum

> Candlesbury (1605): . . . five little bayes built of timber and earth and covered with reed and some straw whereof three bayes have been founded and builded by the now Incumbent, one of the old bayes chambered over with a fastened chamber of a somertree, joists and boards and other old baye having a somertree and joists fastened but the boards loose and moveable . . .'

Not very well-defined specifications perhaps, but builder and owner worked to a common understanding and, apparently, the result was mutual satisfaction.

In the very early days, as far back as the Normans and earlier, there was a hallhouse, a simple cruck frame construction with a central fireplace or hearth and no chimney. Here the family lived and slept with their retainers and animals. The smoke and smell found its way out through the high, unglazed windows, or louvers as they were called. Minor architectural improvements followed through the medieval period. As a general rule these houses are 16 ft, or one pole, in depth, a standard achieved by the simple expedient of sixteen men, gathered together at the church, placing their right feet one behind the other along the length of the village pole. This length was also used to measure a furrow, and became a 'furrow-long' or furlong. In medieval houses this 16-ft measurement is very common: small houses were usually a pole wide and two poles in length—larger houses were two poles in span and four poles in length. In Tudor times the 16-ft pole, or bay, gave way to the 12-ft bay as the first step in conforming with the new Continental method of measurement, the cloth-yard of 3 ft.

With greater refinement in living, more privacy was required, so the master built an upper solar room for himself and his family at one end, with storage space below. In the thirteenth century he began building on another wing, sometimes across the end of the hall, in a T shape. The hall itself eventually shrank, by the use of panelling or wattle-and-daub partitioning, to become the forerunner of our front hall of the present day. The rest of it became the kitchen, with the fireplace moved over to the wall and a chimney built to channel the smoke. As time went on the kitchen, butteries, pantries, etc, were accommodated in additional rooms built on. The storage space became a parlour, flooring was put in to provide an upper storey, and the winding staircase built around the chimney breast took the place of the crude Jacob's ladder by means of which the master had reached his solar.

It is not always easy to trace these improvements and alterations as they have been going on over a period of hundreds of years. But there are a number of clues to watch for, which we will go into later.

## POINTS THAT REALLY MATTER

Now you have had a preliminary look round.

It all appears very depressing. There is worm everywhere, mould on the floor, mould on the brickwork. The ceiling is sagging, huge chunks of plaster have broken away, and the walls have cracks. There are puddles on the floor where the rain has come in through the big gaps in the thatch or tiles, and a depressing smell of dank rot and decay. This, you are sure, is a place that can never be lived in again.

But cheer up. None of these superficial defects mean very much. Go outside and have a look at the roof—don't worry about the holes in it, just see if the ridge line at the top is reasonably straight, with no ugly sags. This will tell you if your main roof members are sound, in which case all you will need are, probably, some new rafters and fresh roofing material—not nearly such an expensive job as replacing the whole lot.

Have a look at the walls outside for any sign of bulging. Even in walls 3 ft thick this can sometimes happen where there is a weakness. It is more common in stone walls, especially those of mortared rubble where the mud used to bind the stones disintegrates, or wet has got in at the top and the stones start to slip. But this again is not fatal; the weak spots can be repaired and, if necessary, the wall buttressed. Cob walls should be on stone foundations, or sills; this was standard practice among old cob builders, who used to say 'give a cob wall a good pair of shoes and a dry hat and it'll last for ever'. The dry hat was of course the thatch with a good overhang.

See if the ground level outside is lower than the floor level inside. If it is not, don't worry. It can be dug away.

Inside the cottage, measure the downstairs ceiling height. Until quite recently it had to be 7 ft 6 in or the cottage had had it, as it is next to impossible to raise the ground-floor ceiling apart from cheating a little by removing the plaster and taking the height up between the joists. This you would want to do anyway in all probability, if the joists are in good condition, but it does not add much more than three or four inches to the room height. Now, however, the building regulations are rather more elastic, and, provided the light and ventilation are 'reasonable' for the room space, there should be no difficulty. If the windows are too small at present to meet these requirements, it should be no great problem to either enlarge them or knock a new one through. It is imperative, especially in cases where an Order has been made on a property, that you keep in close touch with the building inspector and/or his council before you go ahead with any alterations. Building regulations must be considered for all phases of work to do with general construction.

Now for dry rot. This is a fungus, distinct from wet-rot fungus, and can spread insidiously throughout the house. Though it can be eliminated by expensive professional treatment, think twice where you find it. (The sweet sickly odour will often give away its presence.) But if you are very keen do not let it stop you.

As to services, there will almost certainly be water of some sort, probably from a well. Water is the one service man has laid on to his house throughout the ages, and he always built somewhere near a supply of some description: ponds, springs, and streams at first, before he learned to bore down and tap the water underground. It will be there somewhere, and all you need to do is to put in a pump and forget your town water rates. Of course you must have the water analysed, but if the analysis is not satisfactory you can have a small chlorinator installed quite cheaply and you will be 100 per cent safe.

The matter of sewage must be taken into account. If there is no town water laid on there will be no sewers of course, so you will have to provide your own. Modern self-cleansing septic tanks can be located, according to the new building regulations, at a variable distance from the house and nearest water, depending on conditions prevailing, but you must have a certain amount of space to take care of the overflow. Here you will almost surely find the local council surveyor ready to help with suggestions. If may be that, if you are in a water catchment area, a cesspool could be an alternative drainage system.

Electricity is often a problem, especially if the cottage is off the beaten track; but it must be very isolated indeed if you cannot find a line reasonably near. The electricity authorities will usually connect you on a rural development charge, payable by a deposit and quarterly instalments. All the same, it is a good idea to try and find a place with electricity already laid on or at least within handy distance. If the worst comes to the worst and it is out of the question to get a service connected, you can install your own. This does not give maximum power, but for ordinary household use it is quite adequate; and you can get second-hand plants quite reasonably these days, for the electrical power link-up throughout the country is making them

*Page 35* (*above*) Forsaken mill; (*below*) after conversion into roomy double house

*Page 36* (*above*) Ideal prospect for restoration. One and a half acres of garden in woodland area and within fifty yards of electric power line; (*below*) restored derelict farmhouse in the Quantocks

increasingly redundant. An advertisement in the local paper should bring results.

## OBTAINING ADVICE

Now all you have done so far is to satisfy yourself that the property is worth further investigation. Obviously you are not going to trust to your own inexperienced judgment before going ahead and bargaining with the owner; and this is the point when most prospective buyers run unhesitatingly to the nearest surveyor with an impressive chain of letters after his name. Our humble advice is—don't!

He will drive out to your depressed little cottage in a smart new car (that is the average surveyor—we apologise to the others), and spend a considerable time with an array of imposing gadgets, poking, testing, climbing around the roof, knocking nails into the walls to connect to a meter for damp tests, getting his elegant suit covered in cobwebs (for which you will pay, never fear). At the end of it he will depart, looking ominous, with the promise to send in his report. If pressed, he may say, 'Not too good, I'm afraid.' And therein lies the crux of the matter. He *is* afraid. It is not his fault, he has reason to be. And when he sends in his lengthy report, together with his bill for anything from 25 to 50 guineas, it will be the most depressing document you ever read. Because it is his insurance.

According to the law in Britain at present, if you go ahead and buy this property on the recommendation of his report, and if at some time in the future you should have any trouble with some item which he has not listed as being suspect, he, the unfortunate surveyor, will be wholly liable and you will be able to take him to court and sue him for a healthy sum. His only safeguard, therefore, if he is in doubt at all, is to condemn, condemn, condemn. And it is this situation which is to a certain extent responsible for many old buildings being left to rot. It is not a fair situation—not to the properties, not to the buyers, least of all to the surveyors, who often, when approached verbally and 'off the record', will give a very different opinion.

So our suggestion is to take an experienced builder of the old school along to see your cottage. Choose a local man, a modern prototype of the original builders. He will know instinctively as well as professionally what needs to be done, what looks bad but is actually sound. An instance of this was when we were converting an old chantry on the outskirts of a Somerset village. The roof members, sound as a bell, were the weirdest conglomeration of ties, cross-beams, collars, wind-braces and purlins. A surveyor came to have a look at it. 'Good lord,' he muttered as he poked his head through the trapdoor. 'I'll give it a year at the outside. See, the roof is spreading. The walls could go any day.' Now, should a qualified modern surveyor be expected to realise that the roof of a cruck-constructed house cannot possibly spread? The walls will crumble first. Cruck construction is the medieval evolution of the earliest building man put up—two tree-trunks lashed at the top in an inverted V, the ship frame built upside down. Later they shaped or grew the trees, good English oaks, into graceful curves so that when they were set in foundations, wedded at the ridge and secured down to wall height by massive supsport and braces, all interlocked and mortise-pinned

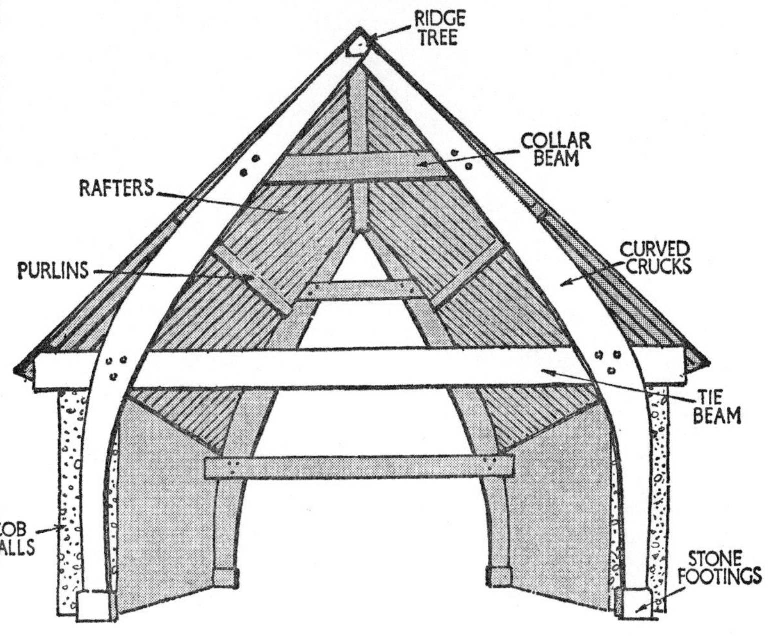


End elevation of early cruck construction

to the solid rafters, there was no chance of the roof spreading.

Cruck members can often be found embedded in cob walls, or clearly defined in stone walls, right to within 3 ft of the ground. In some very early houses they could be traced going right into the ground, as there was a theory that if the tree-trunk were erected root-end upwards the natural flow of the sap would resist drawing up damp from the ground. As it was not always too successful and the base of the cruck had a tendency to rot, they were later placed on a stone ground-sill. It is strange to think that today, when there are minute directives for every phase in the building of a house, when the building regulations are so complex that surveyors have to take a crash course in order to comprehend them, that there is probably more shoddy building going on than at almost any time in history. The houses built with native common sense and pride of work 400 and even 500 years ago may well be standing long after these masterpieces of modern know-how have buckled into oblivion.

So for your old cottage bring in the man who learned all he knows from his father and his grandfather, who in turn learned it from their forefathers. He might not be your first choice if you were going to build a modern villa, but for this job he will give you the best advice and carry out the work in a style that will match what already stands.

CHAPTER 3

# The Purchase and the Plans

---

You have decided to buy. You have seen the owner and agreed a price. If you were buying a ready-to-move-into house the next step would be to put the matter in the hands of a solicitor to let him draw up the contract. But it is advisable in this case to take one intermediate step; and that is to decide on your structural alterations, have plans drawn up, and go with them to the local council. If you are worried about losing the property it would do no harm to go ahead with the contract—signing it subject to the usual searches and subject to your plans being acceptable to the planning authorities.

As regards plans, if all you intend to do is to rearrange the interior, you will need only byelaw permission; in other words, the local council will consider the plans at their next meeting and you should receive their decision immediately afterwards. But if you wish to add to the building more than 1,750 cubic ft of space (measuring from the gable peak to the ground), or make an extension exceeding 10 per cent of the existing construction, you will require planning permission, which is a much lengthier and more involved process, sometimes taking up to six months to sort out. Also, you will need planning permission for any additions that raise the height or extend the front of any part of the existing building.

There is one other little snag worth watching for if you intend altering the exterior of the property: the restriction involving buildings scheduled as being of historic or architectural importance. These buildings are officially listed and the area planning office can tell you at once if yours is included. If it is, and you want to do something drastic like substituting tiles for thatch, or putting in new windows of different design or size, your application will have to go before the Ministry—again taking a great deal of time.

We ran into this difficulty with a charming little cottage we were restoring for a North Country businessman who was retiring to another county farther south. He had bought the property, and his local solicitor had made his searches without revealing the fact that the building—mainly because of its picturesque appearance and setting—was scheduled. Only when we presented our plans did we find out that the exterior alterations he wanted, an extra wing pruned to just under 1,750 cubic ft and a few extra windows, could not be considered locally. Yet the cottage as it stood, with two tiny living rooms, an equally tiny kitchen, and two bedrooms, was far too small for his furniture. The prospect seemed hopeless until we hit upon the idea of ripping out the interior almost entirely. It was wormeaten and had no wood of any quality. The new plans for the interior were passed without difficulty. We bought up all the timber from an old Tudor farmhouse and carted the lot—massive carved beams, door lintels with Gothic arches, studded Tudor doors—up into the hills and installed them in the shell. When the job was finished the cottage looked as though the work had been done hundreds of years ago—every piece added matched the period, only the quality was far superior to that of the original interior—built for a woodcutter. We even put in a deep inglenook fireplace, with seats, all built out of stone and smooth-worn oak. It was one instance where a disappointing start was turned into a most successful conclusion.

For the present purposes we will assume that you wish to convert the interior, and merely improve the exterior. You *can* go direct to an architect and get him to draw up the plans for you, but as the advice given here is for those who wish to spend as little money as possible and do much of the work themselves, this is not essential. You can do the planning yourself, in consultation with your local builder, and merely have the result drawn up by a draughtsman or trainee architect who does this sort of thing privately in his own time. You can even draw up the plans yourself, provided they are clearly done and to scale; but it will probably save a lot of headaches to get someone else to do the drawing. A co-operative building inspector will perhaps meet you on the site and give helpful advice—a verbal OK to suitable ideas and off-the-cuff assistan in the way of getting the drawings done cheaply. advertisement in the local paper could well brin sults. The saving will be in the region of £60 perhaps even more in some cases.

First decide what you want. Draw a rough plan of the cottage as it exists, and when you have carefully worked out what changes and alterations you intend, make another plan detailing each amendment to the same scale as the original. Think well before you finally decide upon any major alteration to the structural layout. For instance, if the staircase seems somewhat cramped, or perhaps winds sharply on a central newel, there is really no valid reason why you should not leave it right where it is and use it. We have become so used to the massive hall stairways of the Victorians that we have come to accept them as standard. Georgian and Regency houses, while having the elegant slender staircase winding out of the hall, are not so aggressive about wasting space and creating draught-funnels. Besides, Regency staircases are architecturally a feature of the hallway.

Where you have a minimum of space to waste and where comfort is your prime object, the cottage stairway cannot be improved upon, especially as it usually has a door at the bottom to shut off the upstairs regions completely. You might, at first glance, consider it dangerous; but consider how many generations of old people and small children have negotiated it safely. Once you have learned to live with your little staircase you will wonder why you ever needed nearly 2,000 cu ft of draughty space just to go up and down. If you feel strongly about having a more conventional staircase it is usually quite a simple matter to have one put in. It is advisable, however, before doing so, to consult your local building inspector. Any necessary alterations to your kitchen, and perhaps the construction of a bathroom, are more or less a matter of individual preference, as long as they can be adapted to the existing layout. Although you should prove capable of doing a large proportion of the restoration work yourself, and with a feeling of confidence, when it comes to such ticklish items as plumbing or electric wiring it is essential to call in the skilled tradesman.

Most old cottages throughout the country have, or had originally, a fairly elementary floor plan, usually two rooms downstairs and two up, both sets of rooms often intercommunicating, with a lean-to at the rear which served as work area and may have been transformed into a kitchen. Of course there are variations and you may be lucky and get something already well planned. The cottage is most unlikely to have a bathroom, and this and your remodelled kitchen will have to be added in such a way as to minimise water-heating and plumbing costs. Suppose you decide to dress up the lean-to or divide off a section of your second room downstairs for your kitchen, then the bathroom should be fitted either in the corner of one of the upstairs rooms nearest the lean-to, or next to the kitchen downstairs (see diagram).

If you want a big living-room in place of two small parlours, the dividing wall can be removed. If your plans make this essential, great care should be taken, and this will be dealt with in Chapter 9. Upstairs you may want a passageway leading to two bedrooms and perhaps a bathroom. This can be done by cheating a few feet off one of your bedrooms (see diagram).

Get to know your cottage intimately. Trace its construction. Its builders were simple people who probably did not know much more about the finer points of building technicalities than you do. They put the walls on a rough foundation, leaving spaces for door and windows. When they came to the second storey,

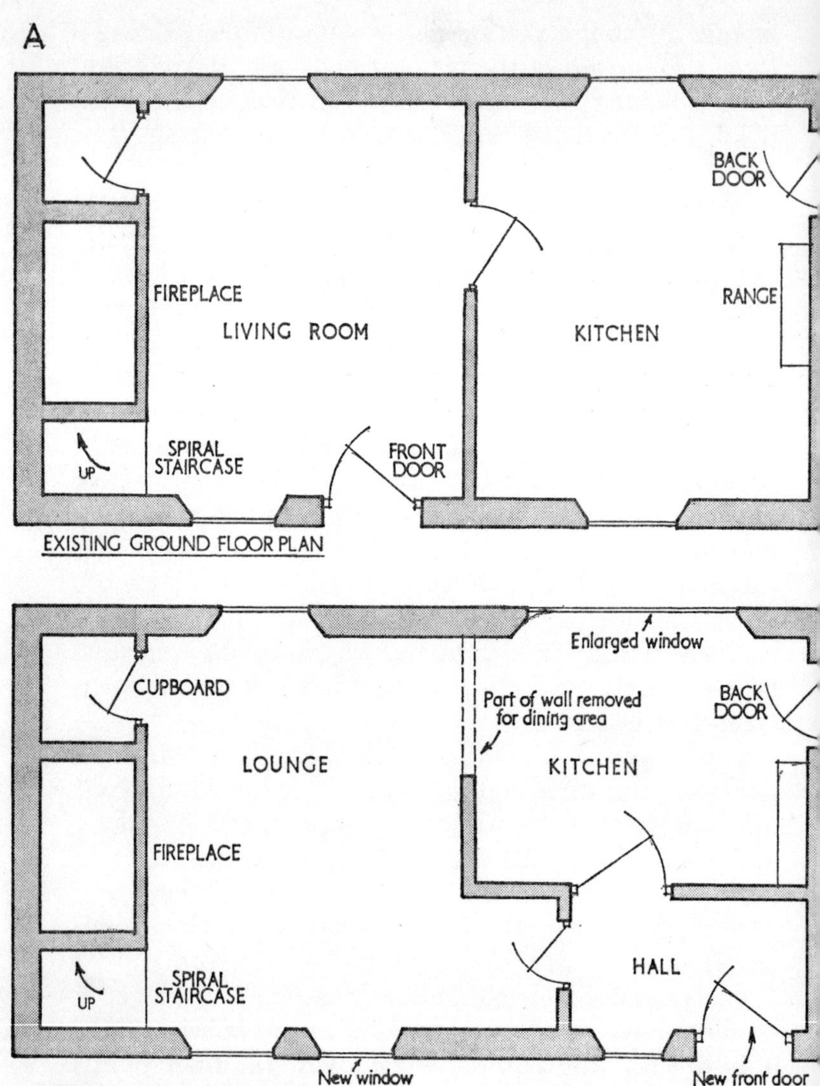


Rough plan showing how simple two-up two-down cottage could be converted

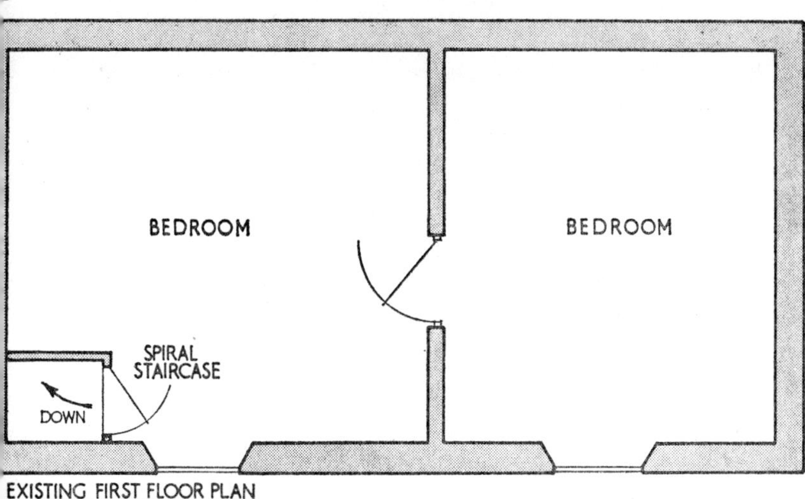


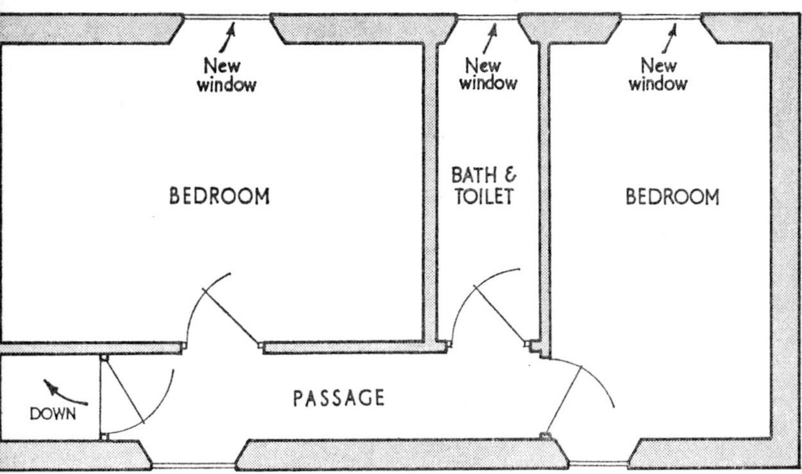


Rough plan showing how simple two-up two-down cottage could be converted

so one old-timer told us, quoting his grandfather, they gathered up all the able men of the village at the site, providing them with a barrel of rough cider, and when they were feeling sufficiently merry, put them to hoisting up the huge summer beams into position on the walls.

Water and drainage will depend on the facilities available. Here you will have to work with your local council regarding the latest building regulations. But you will find they are only too ready to be helpful and give advice in the matter. In the end you will discover that your plans are the product of the concerted efforts of yourself, your builder, and your council surveyor. And if you listen with respect for their knowledge and experience you will save yourself endless trouble and expense.

With the plans drawn, submit as many copies as are necessary, together with the required forms, to your local council. Then sit back and wait. In due course, if you have done your homework, the necessary permission will arrive, with a little sheaf of dockets pertaining to each phase of the work to be carried out. You can then complete your purchase and get ready to start work.

CHAPTER 4

# The Roof

---

You must first decide how much you can do yourself and how much you are going to leave to your builder. Study your plans on the site. Obviously any major structural alteration, like re-siting the staircase, putting in a bathroom, or re-siting the kitchen, is likely to require the services of a professional. Similarly, if you are planning to run a passageway upstairs to provide private access to bedrooms, which, perhaps involves cutting through a tie-beam, here again you will probably need a builder. In short, for anything that disturbs the basic construction of the cottage you will be well advised to seek his help or opinion.

Having decided on this, ask him to give you an estimate for the work. If there is more than one builder in the district, get other estimates, too. But favour the builder who is most familiar with this particular kind of restoration and sympathetic to your desires. Have

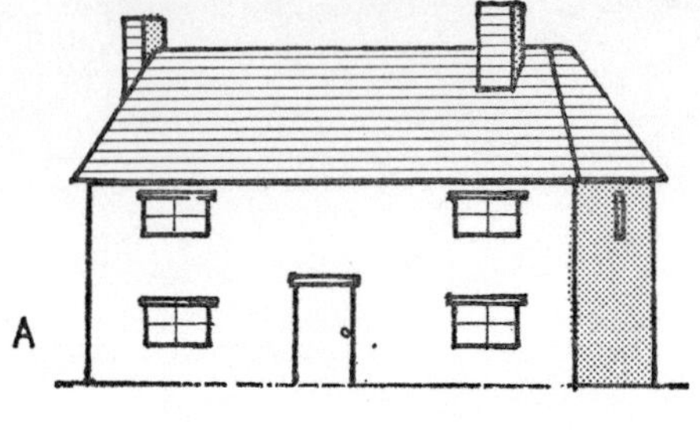


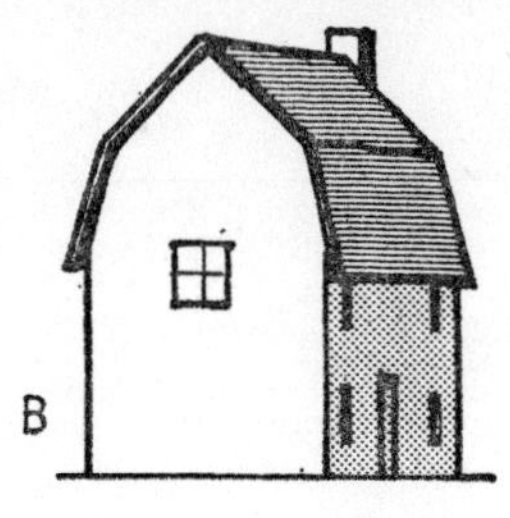

ROOF TYPES

A Hipped
B Mansard
C Queen Anne
D Outshot
E Winged

any major structural work done first, then you will have a clear field for your own operations.

The roof, if you have been fortunate in selecting your cottage, should be well within your province and should be attended to first, as a protection for the interior work you are going to carry out. Similarly, if you have bought your cottage despite a roof that is a complete write-off, get your builder to attend to the re-roofing as a priority.

The particular shape of the roof will depend to some extent on the area in which the cottage is situated. It may be hipped at the ends, or, as is characteristic of East Anglia, mansard-shaped, that is with each side having a double slope. Again it may have an outshot, with one side of the roof extending well beyond the bottom level of the other, a type once popular in Suffolk and the Fen country, though of course also found in many other districts. There is the low-angled Queen Anne style roof, and the valleyed crosswing roof, abutting the main roof at right-angles in the manner of a dormer, common in the Renaissance manor houses and homes of the lesser gentry. But for our purposes we will assume the roof to be a simple straightforward house-covering with pointed end gables.

## TILES

One can almost take it for granted that any house or cottage over 150 years old will be either tiled or thatched. Slates, with the exception of the ancient grey or brown stone variety, were not commonly used as roofing until the end of the eighteenth century, when canal transport enabled them to be distributed more generally and economically, mostly from the blue-slate quarries of Wales and Westmorland. A tiled roof, of flat clay shingles, wavy pantiles, or one of the varieties of Roman tiles, of a neglected cottage will almost certainly have a few tiles missing and others cracked or sufficiently out of place to let in rain. And where water has leaked in over a lengthy period you must expect a certain amount of resultant damage beneath.

But let us, in this first instance, be optimistic and accept that all the wood in the roof structure is sound. The ridge length is reasonably true and not too sagged, so it is just a matter of replacing or resetting some tiles. It is just possible that there is in fact no ridge beam; but unless the cottage is situated in the south-eastern part of England and is of the late medieval period, when it was common to construct roofs by simply pegging pairs of common rafters together at the apex and strengthening them with collar purlins, the chances of there being no ridge timber are remote.

When tiling, always start from the bottom and work up, from left to right, making sure that the nibs of each tile, unless the tiles are pegged, are firmly snugged over the batten it rests on and the lay even. If you are short of a few tiles these should be quite easy to obtain from a neighbouring farmer who can usually be depended on to find an odd stack of them in among the weeds behind an old farm building, and who will let you have all you want at a few pence apiece. Or you can leave the matter for your local builder-handyman.

A word of warning. Don't mix your patterns. Romans have a different 'lay' from pantiles, and unless roof tiles are 'all of a sort' rain, with its evil determination to seek the course of least resistance, is liable to find a way through. Of course it is possible the roof may have been stripped of tiles and slated at some time, in which case any new slates should be set in the same manner as the original ones.

Where the tiles run down to a roof valley and rise

sharply for a dormer or another roof angle, be sure to check that the gulley and filleting or flashing are thoroughly waterproof. This also applies to the fillet around the chimney-stack. And the mention of chimney-stacks reminds us of a few points worth noting carefully. On most old cottages where the chimney-stack is on the side wall or at the gable end you may notice that it tends to lean a few degrees off centre. This is not necessarily a sign of weakness. The old-time builders purposely slanted their stacks towards the roof as a precautionary measure against the elements, so that if a violent wind caused them to collapse they would fall on to the roof rather than crash to the ground below, and be less dangerous to passers-by.

The fact that so many old chimney-stacks are still standing after hundreds of years is fair testimony to the thoroughness of the original construction. It is more than possible that age has caused the mortar between the brickwork or stonework, as the case may be, to crumble away a bit, but a repointing job with a 4–1 cement mix (four parts sand and one part Walcrete) is both fairly cheap and simple to do, providing, of course, you have a good sense of balance and a secure foothold.

Where there is a cowl fixed to the chimney-pot think twice before you yank it off and fling it down into the back garden. While perhaps you might think it unsightly, the previous tenants, or those long before them, may have had good reason to instal it, to suit the prevailing winds and so avoid back-draughts and a smoky lounge. And while up there steeplejacking, check the opening at the top of the chimney-pot. If the cottage has not been lived in for some time it is quite possible that a family of jackdaws, or even several generations of them, may have decided to move in with all their nest furniture and appurtenances; and it is sometimes staggering to discover what these flying junk merchants are able to collect and store. We cleaned out the chimney-pot of an old forsaken cottage and took away several baskets of litter, mostly sticks, but including pieces of string, broken glass, tin-lids, bones, buttons and, believe it or not, an old pair of wire-framed spectacles. As a rule a sweep's brush will clear the nests in a chimney from below, but in some instances the birds will artfully dislodge a loose brick or stone in the top as a ledge to settle on; in which case it can only be cleared out from above.

Note also the angle of the roof. If it is very steep, more than 40–45 degrees, then you can be fairly certain that originally it was thatched. To establish further proof of this see if there is a stone-slab coping down the gable ridges approximately a foot above the level of the tiles, and evidence of filleting or flashing around the chimney-stack at the same height. If so, that marks the original level of the top of the thatching.

## THATCH

So much for tiled roofs. With thatch, regardless of its conditions, there is little you can do personally; it is strictly a job for the expert. The life of an average thatch job, with rye or wheat straw, is about twenty to thirty years; with Norfolk reed probably twice as long for twice the price. So if you have to have a rethatch our advice is to settle for the local material. This should cost in the region of £400–£500, depending of course on the area of the roof, for a complete skinning and rethatching. If the thatch is in fairly good repair with, let's say, only one or two worn or shaggy spots, then a good thatcher will recommend simply making good the bad patches to conform with the existing thatch. The new bright-yellow straw may stand out like a sore thumb, but it will soon weather and blend in to become unnoticeable. And your pocket-book will heave a sigh of relief.

Some owners of thatched cottages insist on having their roofs covered with a tight fine-wire mesh to keep birds from pecking holes in the straw. While there may be something to be said for this, at the same time it more often defeats its own purpose of preserving the thatch by thwarting its natural tendency to ward off water. The elementary water-shedding principle of straw or reed thatch is to allow the water to run unrestricted from straw to straw down the steep angle of the roof and drop off the eaves before it has time to soak through. On sound thatch, even after weeks of continuous rain, it will be found that wet has not penetrated much more than an inch below the surface. So with a blanket of tight, well combed straw, 8–14 in thick, the chance of a leaky roof is, to say the least, remote.

But when a thatch is cased in with wire mesh the flow of rainwater has to hurdle each horizontal wire on its way down the roof, thus giving it more opportunity to soak down through the surface. And, more convincing still, is the argument against mesh put forth by one of the finest craftsmen in the West Country, who came from a family of thatchers, father and son, going back eight generations. 'Do 'ee put wire-mesh atop o' thack 'ee spoil 'un. The wind do whip the straw agin the wire an' breaks the ends. Now thick end 'e cassn't blow away on account o' being held down by the wire, y'zee, so 'e turns down in. Then when it do rain the water runs down the straw deep into the lay and purty zoon the wet straw begins to rot underneath. 'Tis a poor way to treat a good thack, an' all agin common zense.' So on balance it is better to leave well enough alone, even if plagued with a few straw-hungry birds.

You will probably have to pay a higher insurance premium for a thatched house than you would for one with a tiled roof; for, according to most insurance companies, there is a greater fire risk with a thatched roof. Although this may be true up to a point, it is well to remember that many more old thatched cottages have been pulled down than have been burnt down. There was a law in medieval times that required all thatched roofs to be whitewashed so as to make them 'not so ready to burn'. In parts of Scotland, Yorkshire, and mid-Wales, the custom has survived the obsolete law; and in the Cardigan Bay area not only white-washed thatch is still to be seen, but whitewashed slates also. Today it is possible to spray thatch with a chemical fire-proofing liquid which, although not wholly reliable, especially over a length of time, will go a long way to allay fears of a roof easily catching fire.

## SHINGLES AND TILE-HANGING

Shingling, once quite a common form of roofing, has now almost passed out of existence, though there are still a few old cottages and barns with shingled roofs to be found in rural England. The 'split' or 'cleft' shingle, a sliver of wood about 8 in × 12 in, and overlapped on the laths, very likely preceded the use of stone slates and tiles. It is on record that in the year 1260 King

Henry III ordered the thatch to be removed from an outer chamber in the tower of Marlborough Castle and replaced with shingles. Wooden shingles were more commonly used on important buildings during the early Middle Ages; Salisbury Cathedral was originally roofed with shingles from the New Forest. But although wooden shingles served their purpose well and were attractive to look at, they had two large disadvantages: they were expensive and their life was comparatively short. By the beginning of the fourteenth century the shingled roofs of so many manor houses, church steeples and castles had fallen into such bad repair that it was found they could be replaced at less cost with stone slates laid on moss, or with earthen tiles. The use of wooden shingles considerably diminished by the end of the century, though it was continued in some south-eastern counties, principally on church spires.

The early English settlers in America found shingles cut from cedar or redwood a most convenient form of roofing for almost all their log or weatherboard buildings. Today they are still widely used there, and shingle mills are kept busy sawing them in 16-in lengths and random widths.

Tile-hanging was another old craft, once employed extensively on cottages in the southern counties, particularly in Kent, Sussex, Surrey, Hampshire and Berkshire. Unfortunately, few examples still exist. Tile-hanging appears to have first become popular during the Jacobean period, when the increasing demand for tiles necessitated their local manufacture rather than reliance on supplies imported from the Continent. In the majority of cases tiles were hung on the gable-end wall of the first floor, where the timber studding had perhaps rotted or shrunk, allowing damp to penetrate gaps left in the infilling. Later, from what was originally a practical means of weather-proofing an outside wall, there developed a sudden craze for larger expanses of tiles. Many houses and cottages had all four walls hung with them solely for reasons of ornamentation.

Tiles used for wall-hanging were specially made for the purpose. They were flatter and thinner than the heavy roof tiles, and were lapped and fastened over oak laths with pins of hazel or willow. Sometimes they were bedded in mortar, sometimes in damp moss. The old workmen took great pride in handling them, and developed an instinct for matching the right textures and colours, an outstanding characteristic of hanging tiles. The soft tones of red, brown and purple hues, after years of exposure, give a most picturesque appearance to tile-hung cottages. A considerable variety of patterns was made: besides the plain rectangular tile, there was the oval-bottomed type, the double-bow, the flanged tile and many others. Sometimes the tiler would mix his patterns with pleasing contrasts.

At the time tile-hanging was popular in the south-eastern counties, the western side of the country, including Wales, had turned to weather-proofing outside walls with indigenous slates. A vast number of cottages had their walls slate-hung from top to bottom, the silver-blue slates often arranged in attractively varied patterns of horizonals and diagonals, giving an overall effect of neatness and comfort.

During the Victorian period an attempt was made to re-introduce tile-hanging; but, like many Victorian architectural innovations, the result was a baroque imitation which lacked all the charm of the original.

## THE ROOF TIMBERS

The roof has so far been approached as a straightforward job, that is with the assumption that all was well directly below—sound battens, sound purlins and rafters, with cross-braces, collar supports, tie-beams, and main members all giving 100 per cent service. But, frankly, this is too much to expect. Time and nature, in their own interests, have probably left their mark. So let us reappraise the condition of the roof and prepare for the worst.

Lift off a square yard or so of tiles at a few random spots along the roof to let in the light and climb down on to the joists below. In the case of a thatch covering, particularly if it is sound, this method of entry is of course out of the question, and you will have to reach the same point from the inside. There may be a trapdoor in the upper ceiling, though as a rule old cottages never bothered with access to the roof from within. Where there is no trapdoor one will have to be made. Choose the most convenient place in the ceiling for a roof access, if possible at the top of the stairs or in the upper passageway, and cut a hole large enough to climb through in the plaster and laths, between two joists. (It is a simple matter to frame the opening with two lengths of 2 × 3-in wood nailed to the joists at each end, and to make a lid that will drop snugly into position, to be plastered in to a smooth finish and decorated later.)

A good probing kit to have with you on these surveying expeditions consists of a stout penknife with a long blade, a small sharp hand-axe, a stiff wire brush and an electric torch. It is just as well to come dressed for the occasion also. You are about to disturb perhaps centuries of accumulated dust, straw, and spiders' webs, so wear the oldest clothes you have, with a neckerchief and by all means a hat well pulled down.

The old method of roof construction generally used almost regardless of the quality of the cottage, relied on the common principle of supporting the weight above and holding rigidly to the walls on all sides. The actual measurements and trimming of timbers were not considered important so long as the proportions of their various bulks were fairly consistent. Only in manor houses and some of the wealthier yeomen's hallhouses, among medieval secular buildings, where all the interior timber framing was exposed from floor to ridge, did the carpenter bother to exercise his skill: here were carefully matched smooth-adzed beams, rafters, and even kingposts, each chamfered, parts ornately decorated with fine carving, and all of oak-heart or deep-cut elm. In some of the elaborately timber-framed houses in the lower Midlands and the eastern counties, especially where the upper storeys are jettied or overhung, the exposed black-oak studding, lintels and ends of the huge dragon-beams show evidence of the woodcarver's patience and skill.

In a cottage, however, the local carpenter who put it together was working strictly on a utility basis. You will find that the rafters are probably crude pit-sawn lengths with here and there a near-enough matching branch with the bark still clinging to it. The cruck blades or truss joints which come together at the ridge, joists, and purlins, are just as rough-hewn as the rafters, though stouter. The batten strips running across and on top of the rafters to hold the tiles (or spurs, in the case of thatch), look too flimsy to do the work. Still, they haven't done so badly over the years, and most of them are good for a long time yet.

But to get to work. Pick your way carefully over the joists so as not to put a foot through the ceiling, and examine the woodwork closely. Brush away the dust and cobwebs where it may look wormy or show signs of rot, particularly at the jointings, and test any damage by digging in your knife. There may be a beam that looks wormeaten all the way through: chop or scrape away until you come to a firm surface. You may have to hack off a lot of soft wood before you have done, but don't be too discouraged. As long as the beam is sound through the middle without lessening the diameter too much, its bearing strength should not be affected.

In one old Tudor house we restored we found rot and worm in a massive oak bressummer beam which ran the entire width of the house, over 24 ft, and was the main support for the chimney-breast and two upstairs rooms. The surveyor who accompanied us dug in with his knife, which sank almost to the hilt. He shook his professional head and pontificated, 'That will have to come out. It's a wonder it hasn't collapsed and brought the house down with it long before this.' But the old-time carpenter we had working for us just smiled. After the surveyor had gone, he took up his hand-adze and set to work chipping away along the full length of the beam until he came down to the hard wood, almost 2 in below. After a good smoothing off with his power-sander and the application of a coat of tinted preservative, the beam came up shiny black with a surface as hard as iron. ' 'im and his book learnin',' the carpenter said, standing back to admire the bressummer, which looked as natural and as beautiful as ever. 'So now 'tis only fourteen inches through instead of sixteen. It still be stout enough to take the weight of another two storeys on top and hold 'em steady through an earthquake.'

Of course, where some of the roofing timbers may have been long exposed to damp or dripping rainwater or serious infestation there is bound to be a certain amount to write off completely, especially in cases of rafters and purlins. Rip out all the wood sections that are too far gone and replace them with new material, being sure to treat it first with a reliable preservative.

It is always well to bear in mind when inspecting the timber in an old house or cottage, that what you find is largely a matter of luck and local conditions. Once you have satisfied yourself that the woodwork is in good repair it is up to you to see that it stays that way. So with an old bucket and brush and a good supply of wood preservative, go over everything you can reach, after first brushing away as much dust as you can. And don't be sparing: make certain the liquid gets well into the joints and wormholes. If the weather is dry at the time it is advisable to remove as many tiles from the battens as possible while you work, both for light and for ventilation. In the case of a thatched roof where your ventilation might necessarily be restricted, do only a section at a time and let the preservative dry off a bit before continuing. Once the inside of the roof is made completely sound, and both wormproof and damp-proof, proceed with its outside, as explained earlier. Then you can reseal it with the confidence that it should remain sound for a long time to come.

CHAPTER 5

# The Exterior

So much for the roof. Now let us look around the outside of the cottage.

## WALLS

Instead of a nice tight angle joint at the foot of the wall and the ground, there is the almost inevitable flowerbed heaped up against the wall, with climbing roses, creeper and probably ivy, planted long ago and left to grow in wild abandon. While this luxurious growth does much to enhance the quaintness of 'ye olde-worlde English cottage', it also has a deleterious effect on the walls it so charmingly decorates. So our firm advice is, out with it—root and branch! Yes, we know what it is like to have to tear out colourful Virginia creeper, red-berried pyracantha, and even espaliered fruit trees: it hurts. But if the choice is either to put up with damp crumbling walls or to move the garden edge back 6 ft or so, the decision shouldn't be difficult to make. Remember, we want a cottage that is sound and comfortable, when it comes to priorities—one that will do justice to the work you put into it.

With the walls free of horticultural infestation—particularly ivy, that haven for so many insect pests, whose tentacles have a nasty habit of searching for a crack and working their way through to expand and increase the damage—we now have an opportunity to study their condition and make any necessary repairs.

If the walls are built of solid cob, they will be anything from 20 in to 3 ft thick. They may be of a cob and stone aggregate, or solid rough stone with lime and mortar or clay binding. Or, as is quite common in soft-stone districts such as the Ilminster area in Somerset, they may be of ashlar, square blocks of cut hamstone on the outside facings, rubble-filled. In the Norfolk and Suffolk areas many smaller houses and cottages, built before the ready availability of brick, were timber-framed with flint base-courses and in-filled with lime plaster or flint-faced rubble, the lime being excavated from the local chalk pits. In some cases the flints were knapped to expose their characteristic black face. Sometimes whole pebbles were used as a facing for cottage walls. In Cambridgeshire there are still to be seen many old cottages constructed of clay lumps, which were pressed like bricks but unburnt. Further west the use of sandstone was not uncommon for walls, particularly in and around Warwickshire. Along the Cotswold scarp crude stone rubble was the predominant building material, whereas to the north, up through Worcestershire, the half-timbered sixteenth and seventeenth-century dwellings were infilled with wattle and daub. In areas around Northumberland, in fact extending well over the border into Scotland, simple cottages were constructed of turves, sometimes 6 ft thick and faced with dry stone, a method of building was known as 'spetchel walling'.

Each particular district of course made use of the building material most readily at hand, although until the time of the Renaissance solid stone was generally used only for foundations or groundsills in ordinary English buildings. Even in south Yorkshire, Pembrokeshire, Cumberland, and Cornwall, where good building stone was plentiful and easy to hand, almost all secular common constructions were built with a wooden framework, clay or clom daubed, unless they were fashioned of solid cob. It is interesting to note in passing that in the year 1212 ordinary dwellings in London were so flimsily constructed that the City Aldermen were provided with 'crook and stick' to pull them down when they caught fire.

Brick, meaning burnt brick as opposed to sun-dried clay lumps, was in many parts of the country little used as a building material until well into the seventeenth century, though the Romans were using lacing courses of cemented bricks in Britain at the beginning of the Christian era. These Roman bricks were flat squares an inch or so thick, and would today be called heavy tiles. In some of the eastern counties, however, bricks imported from Denmark and Holland were being used a century or so before they became generally popular over the rest of the country. It is difficult to understand why the employment of bricks for ordinary buildings was so slow in spreading, unless it was because they were too expensive and because there were not sufficient 'brycke-layers' to go round. The West Country was the last area to adopt them: the first whole building of burnt brick in the western counties, Gray's Almshouses in Taunton, built in 1637, is still to be seen.

It took the Great Fire of London in 1666 to give the much needed fillip to Britain's brick industry; in the rebuilding of London, timber was replaced by burnt brick, Sir Christopher Wren being on hand to supervise the use of this 'new' durable and fireproof material.

The size of bricks has altered little since the time of Edward IV. In 1477 it was declared by statute that the measurements of a burnt brick should be $8\frac{1}{2} \times 4 \times 2\frac{1}{2}$ in, the reason given being that this was the most convenient size and shape for a bricklayer to grasp with one hand. Early bricks, most of them made by Flemish brickmakers from the mid-sixteenth to the early seventeenth centuries, can be distinguished by their peculiarly rough texture and somewhat irregular dimensions. The mortar joints in early brickwork are coarse and the wall thickness is usually a brick and a half, or about 14 in. At first bricks were laid with little attention paid to bonding patterns. Much of the charm of old brickwork, especially that of the Queen Anne period, is due to the irregular bond as well as to the rich colour of the hand-rubbed face bricks. The bricklayer considered the wall to be satisfactory so long as the vertical and horizontal lines were true, and the coldly mathematical precision of modern brick building construction was absent. The bricks were mortared tight together through the wall thickness, so that the mingled headers and stretchers interlocked to

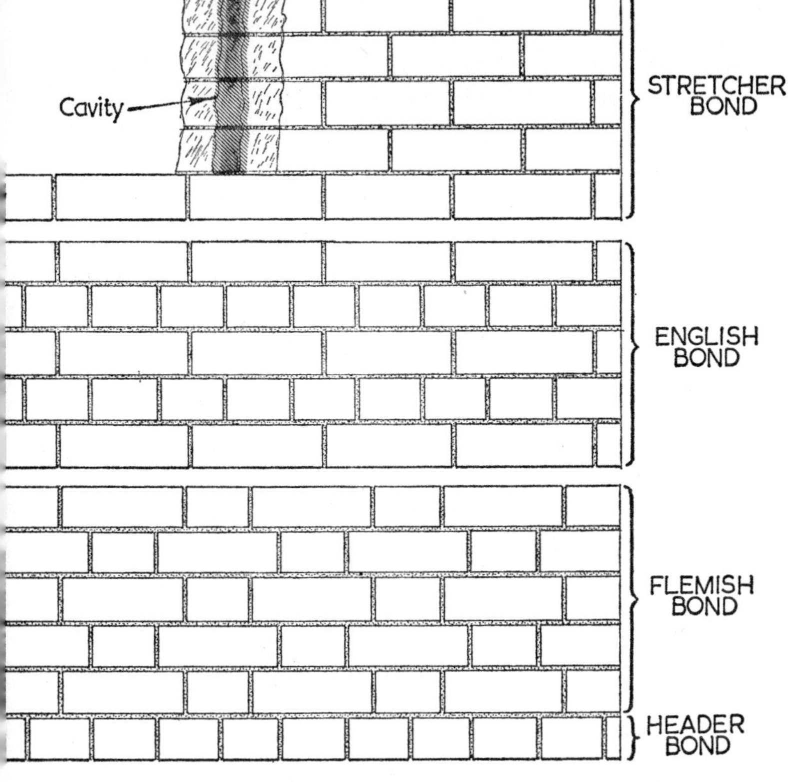


Diagram showing most common brick bonds

ensure a solid bulk. Later, the more experienced bricklayers adopted a systematised pattern for walling which came to be known as 'English bond', alternate courses, or layers, of headers and stretchers. English bond was supplanted by 'Flemish bond', in which headers and stretchers alternate in every course, during the seventeenth century. Not all bricklayers conformed to the accepted patterns of bonding. Some preferred to make patterns of their own—heading bonds, stretcher bonds, in which every brick was laid either head or length showing on the wall face. There was the 'Garden wall bond', three courses of stretchers and one of headers, the zig-zag 'Herring-bone bond', used particularly for infilling timber-framed houses, and various other designs.

Modern brick walls are constructed to a standard of two 'skins', or single-brick widths, of $4\frac{1}{2}$-in stretcher

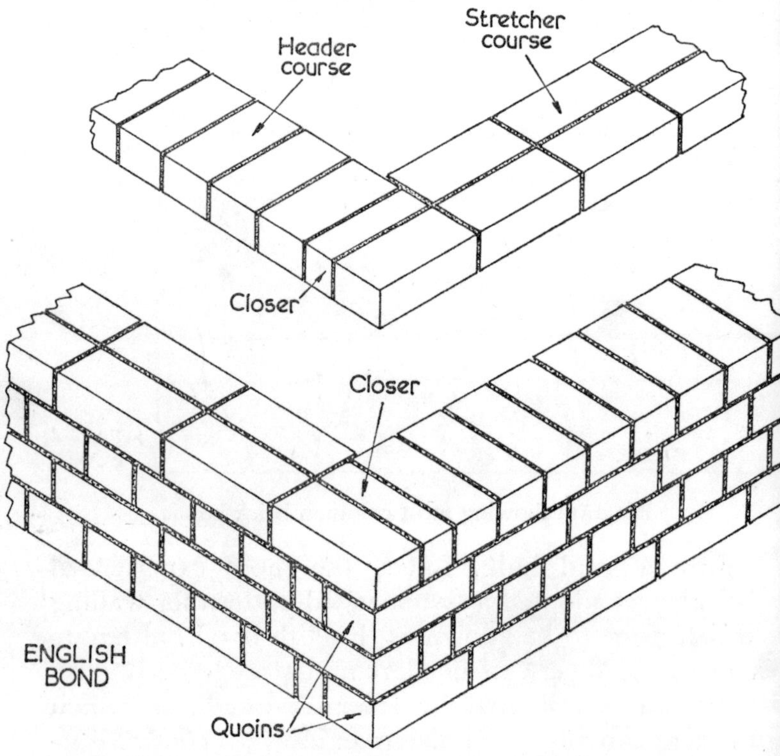


Nine-inch wall in English bond, showing how the corner is formed by using 'closers' (2-in cut bricks) to break the vertical joints

*Page 69* (*above*) Sixteenth-century jettied Warwickshire house (note brick nogging infilling); (*below*) hall of late thirteenth-century Somerset manor house showing roof construction. The plaster has fallen away from the area which was once the louver or smoke hole in the ceiling

*Page 70* (*above*) Example of fourteenth-century cruck construction. The oak pegs through the collar-beam are original; (*below*) exposed cruck and beams in bedroom

bond with a 2-in cavity between. The two thin walls are linked for stability by metal wall-ties spaced at regular intervals as the bricks are laid. The cavity wall provides air-space insulation against both sound and damp penetration.

Throughout the 'highland zone' of Great Britain (that part of the country roughly north-west of a line drawn from the Tees to Torquay), where stone was abundant, stone walls were bonded almost entirely with common mortar, ie earth beaten up with water without any admixture of lime, as recently as the late eighteenth century. Clay and cowdung were sometimes used as additives. It would seem that the early builders knew nothing of the chemistry of mortars, in spite of the fact that in Roman Britain stone walls were bonded with a mortar of earth and lime. In some districts walls fashioned by the simple method of piling stone upon stone without any mortar bedding whatsoever can still be seen. As the wall grew in height so it was plastered on both sides, the stones being relied upon to settle firmly in position by virtue of their own weight and what support they could be given by interior timber bracing.

Cob construction was better understood, and anyone wishing to treat and repair cob walls properly should become familiar with it. Cob, or 'clobbe' as it was known in the Middle Ages, was a mixture of mud, clay —if it was available—and straw. So a house made of good cob, especially if local chalk or lime had been added, could be said to be built entirely of reinforced mortar. The method of mixing varied slightly according to the material at hand. Cob architecture was used in Babylon thousands of years ago, and is still used in some parts of the world today; in Mexico many settlements are built entirely of mud, or adobe as it is called.

The mud, with or without the other additives, was beaten to a sloppy consistency and allowed to stand for a day or so to 'temper'. This procedure was often repeated until it was deemed satisfactory for use, and in East Anglia and Cheshire horses trod the clay and straw into a 'pug'. When the cob was ready the first layer was built all round the foundations of straw-covered stone to a height of 24 in. While it was still wet a layer of straw was put on top, with the ends pointing outwards, and well trodden down by the cob masons. It was then left a week or so, depending on the atmosphere, to dry and harden, when the next layer was applied; and so on, each layer being strawed and trodden down in turn. Time was usually not the prime factor in these operations, probably just as well for it often took as long as two years for the walls of a two-storeyed house or cottage to be properly 'raised'. Woodwork, such as beams and lintels, was fixed in place as the work went along. The walls were trimmed inside and outside, plastered or daubed with a well-beaten mixture of clay, cow dung, and finely chopped straw, sometimes with fine gravel added, and then coated with lime-wash.

The repair of cob walls is not difficult, providing the right material is used, so some cracks and flaking need not condemn your cottage. You will find you can do the work yourself quite satisfactorily, and in fact by virtue of being an amateur you will leave a natural rough finish more in keeping with the original than a master mason would do. Remember, old-time rural builders worked by rule-of-thumb. They had no architects plans to consider.

When we first started doing restoration work we called in a bricklayer to rebuild a 6-ft gap which had once housed a pair of loft doors in an upstairs wall. The original wall on either side of the gap leaned in 2 in at the top and was so 'out of true' as to be distinctly bent. The young bricklayer, who had learned his trade in an up-to-date technical school and whose sole experience as a journeyman was taken up with building modern houses, did a first-class job. In fact, he built the wall with such plumb-line and spirit-level perfection that it had to be torn down and rebuilt by a local handyman-gardener, who made a good rough job with old bricks and followed the original contours so as to blend in with the existing wall line. Such is the difference between restoration and renovation.

Starting from the top, pry off any patches of the plaster rendering that bulge clear of the cob wall and also any loose flakings you find. Don't be shocked if after you have pulled off two or three wheelbarrow-loads of plaster the cottage looks as if it has been blitzed and is ready for the demolition gang. Old houses have a way of looking their worst when they are being prepared for restoration. It is vitally important to use the right mix when marrying in to old mortar or on to cob. Modern cement was unknown before the first quarter of the last century, and, when mixed with sand, dries with a brittle stone-like finish quite unlike the soft porous cob. Consequently, when subjected to temperature changes, the expansion or contraction causes the hard cement to pull away from its base, and so invites the wet to seep through. There is on the market, particularly in builders' supply yards, a soft-setting cement known as Walcrete. This is much more suitable for the purpose, as it contains a percentage of limc. One part Walcrete mixed with four parts soft sand will bond in with the cob and ensure a good seal. Or if Walcrete is unobtainable, make a mix of one part cement, four parts soft sand and one part lime. Plaster the mix over the bare patches, making sure you work it well into the edges all round. Trowel it to an even thickness to blend in flush with the wall. When plastering round door and window lintels, exposed studding, or other timberwork, first pry out any cracked or crumbling cob before you make good.

With old exposed brickwork it is quite possible that the pointing has worn away, leaving deep gaps between the bricks. Test the mortar by scraping it with the point of your trowel. If it is a weak aggregate of lime and sand, in which case it will crumble easily, repoint with Walcrete or cement with lime added. On the other hand you may find that the bricks have been repointed comparatively recently by someone unfamiliar with complementary materials who has used hard cement; in consequence the strips of cement have broken away from their seating to leave the brickwork vulnerable for damp to seep through the mortar underneath and so penetrate the wall. In this case take off all the loose pointing and renew with lime mortar.

While on the subject of exterior cement work, be sure not to use cement if there is any danger of frost before it has had time to harden, unless you include in your mix an anti-freeze additive. A freezing temperature will affect untreated wet cement and 'kill' its setting properties.

The bond of old stonework may be of lime and sand mortar or perhaps common mud plaster. Repoint the stones as with old bricks, first roughing the mortar to make a key for a good bonding.

To prevent rising damp in brick or stone walls it is essential to have a damp-proof course just below the inside floor level. Only in comparatively recent times have damp-courses been made compulsory—although some sixteenth-century manor houses in fact possess built-in damp courses of inch-thick slate. A damp-course can sometimes be inserted by the amateur, applying the old-fashioned method of taking out a few bricks at a time, laying in damp-course felt or a plastic barrier, and cementing the bricks back on top of it; and proceeding in a like manner all round the house. But such an operation can be tedious for the inexperienced and is often unsatisfactory, particularly if there is uneven stonework. A more dependable solution to the problem is to put the matter in the hands of a specialist, who will apply a chemical injection of silicone and guarantee results lasting up to twenty years. Brick walls can be rendered completely waterproof by treating them with a special silicone fluid, easily obtainable from dealers.

## GUTTERING

The guttering, unless the cottage is thatched, in which case it will not be needed, is something that requires special attention. Even though the fitments and downpipes may be in good condition, the use of a long ladder and a garden trowel will be found almost essential to clear the gutters of rotted leaves and mud and to free any choked run-offs. Also check the guttering-boards, gutter joints, and brackets. A loose bracket can disrupt the angle of flow and is often responsible for constant dripping.

If the guttering needs repair consider well before you attempt it yourself. The chances are that it is the old-fashioned cast-iron variety, which your hardware merchant may not be able to match. Most guttering these days is made of galvanised iron, asbestos or plastic. Any of these is suitable as a complete set, but joining in a separate length of alien material can be a tricky matter, especially if the thickness varies enough to interfere with the natural run. The galvanised-iron guttering would probably suit best, as it is fairly simple to cut with a pair of tin-snips and to bend to shape. If the joints don't make a perfectly snug fit apply a sealing compound, easily obtainable from your hardware shop.

At the foot of the downpipe, assuming you are established in a sequestered rural area without main drainage, you will most likely find that the water runs into a soakaway, unless it is carried to a septic tank. The soakaway consists of a deep gravel or rubble-filled hole which enables the water to soak down through and disperse into the subsoil. Some downpipes serve only to keep the water-butt supplied, and it is convenient to have a constant supply of rain-water; but it is important to see that any overflow gets away freely from the side of the house.

The ground immediately surrounding the cottage should be level with the bottom of the walls and below the floor level inside, with an angled run-off so that water has no chance to collect in puddles to soak down under the foundations and be drawn up through the inside of the wall by the dry warmth of the rooms, resulting in perpetual damp. The early builders, and it is a puzzle to understand why, did nothing to prevent rising damp. Although they showed a vast knowledge of construction principles, were able to cater for stresses and strains, and could build a watertight roof, for some peculiar reason they paid little or no heed to the fact that water finds it almost as easy to seep up as to seep down. Perhaps they relied too much on the law of gravity, or perhaps they reckoned that the heat thrown out by the continually burning fire on the hearth would be sufficient to dry off any pervading moisture. Whatever their reason it shows a sad lack of understanding.

## FOOTINGS

With the roof and outside walls having received some attention, let us turn to the footings and their immediate surroundings. As was pointed out earlier, it is vitally important to see that no water has a chance to collect. Unless the previous occupant had taken care of this matter—and if luxuriant growth had been left to flourish this is most unlikely—it is up to you to do so.

The initial job is to level off the ground in contact with the base of the walls all around the cottage to a width of at least 3 ft, allowing for just a sufficient slope for the water to run towards the lowest corner to the drain or soakaway.

Having done that make a skirting round the walls. Prepare a good straight plank, across the ends of which have been nailed pieces of 1½-in batten, and set it along the base of the wall at ground level so that the batten pieces allow a gap. (You may find it easier to use two or three planks of varying lengths, also with battens across the ends.) A few bricks stacked against the plank along its length will keep it in place. Mix up a batch of fairly sloppy concrete—three parts gravel, two sand, one Walcrete—and fill the gap to the level of the top of the plank, poking it well down with your trowel. Set the next plank in position and repeat the procedure all round the house. If your planks are limited wait twenty-four hours for the concrete to set before you take the first plank away. You may find a vent grating on the same level; in this case simply cut a piece of board the same size and 'blind' the vent until the concrete has set. As soon as the concrete has hardened sufficiently for the planks to be removed make good any flaws or bad joints in the skirting all round.

Now make a base, or apron, by concreting out from the base of the skirting, extending a couple of feet to a thickness of 3 in. Trowel up the edges slightly so that the apron is 'saucered' for the water to run. Use a spirit-level, laid on a length of straight batten, to check the slope. Then go around the apron with Walcrete and seal the top of the skirting to the wall, trowelling it off at an angle of 45 degrees. At the same time make sure that the bottom of the skirting is trowelled smoothly to the apron, to ensure a watertight seal.

## WINDOWS

With this work completed, there is just one other item to deal with on the outside, apart from the decorating, before the cottage is ready to do battle with that arch enemy, damp, and hold its own for years to come. This is the woodwork—exposed beams, doors, lintels, windows, studs, and perhaps sills, though few rural building over 150 years old seemed to bother with wooden sills as such.

The existing windows of old cottages often belie the true age of the building, for the original jambs and frames, unless they were fashioned of oak-heart, will have long ago succumbed to the wear and tear of the elements and will have been replaced, perhaps several times. (The word window is derived from the old Norse word 'windauga', that is, wind-eye. Later the Anglo-Saxon term became 'wind-dur' or wind-door, since corrupted to 'window'.)

Early window openings were usually just small holes in the wall, closed by a movable shutter-board or louver. It wasn't until the thirteenth century that wooden frames, made of separate pieces of wood, came into use. To keep out the rain, window openings were latticed by crossing laths or wattle diagonally to enable the water to run downwards. Later, when lead-glazed windows took the place of open wattle lattice, the diagonal crossing pattern was followed for the lead cames. Hence the origin of diamond-shaped panes, or 'quarries', as they were called.

The use of glass filling for window frames of ordinary buildings is comparatively recent. Until the seventeenth century glazed windows were rare, except in churches and gentlemen's houses. The average dwelling found it sufficient to fill in the latticework with thin horn, or hang oiled linen over the frame, to keep out the draught and at the same time allow some light to come in. There are some Elizabethan windows, made of heavy oak with chamfered mullions, still to be seen today in old barns and eave-lofts.

It was around the early Elizabethan period that hinged iron casements were introduced. But these were expensive and only the larger houses could afford them. Modern double-hung sash windows, with weight and line, came into general use in the better houses at the beginning of the eighteenth century, though there were unweighted sash windows, whereby the lower half was raised by hand and pegged into place or slid sideways, in existence well over 100 years earlier.

Traces of window outlines are often noticeable in the walls of old houses and cottages. The chief reason for their having been blocked up was the Window Tax of 1697, when a householder was taxed according to the number of windows he possessed. This tax lapsed for a time, but was reimposed later and not finally repealed until 1851.

The windows of our cottage will probably follow the pattern of most small country dwellings and be four- or six-paned wooden casements with one half opening, usually against the prevailing wind. It can be pretty well taken for granted that, on inspection, any paintwork still showing is in a sorry condition and the putty around the panes dried out and cracked. Restoring badly neglected windows can be a tedious job. But if they can be saved it is well worth the effort.

With a scraper and small chisel go over all the woodwork of the casement, frames and glazing bars, and remove all vestiges of flaking paint and loose putty; then give it all a good sanding until you are left with a more or less smooth finish. If there is any rot, gouge out as much as you can. The two most vulnerable places for a window are where the hinge screws may have rusted and rotted the surrounding area of the jambs or frame, and the corner mortise and tenon joints. Take out the whole casement if the hinges have lost their seating and, after digging out all the soft wood, prepare a plug of batten a little wider and longer than the hinge with a couple of 1-in screws fixed half-way into the back. Treat the cavity with a fungicidal fluid to arrest any rot which might otherwise tend to develop owing to the damp already present, and fill it with a cellulose wood filler. While the filler is still wet press the plug, with the screws embedded on the inside so they will ensure a firm hold when the filler hardens, and dress the surface to a smooth finish, flush with the jamb or frame as the case may be. If the corners of the window are weak, screw a flat 2-in angle bracket to both sides, first cutting a rebate so they seat flush with the frame.

As soon as the cellulose filler has set hard you can refit the window, remembering to chisel a rebate in the plug to allow for the thickness of the hinge. You may find you will have to plane or sand off the edges of the opening window to make a comfortable fitting. With all the old cracked and loose putty prised off, re-putty where necessary, pressing firmly down with your putty-knife, and fill any nailholes or splits in the wood-work at the same time.

Use a stiff wire brush on all exposed beams, lintels and other timbers, and treat them liberally with in-secticide, either clear or coloured, after, of course, you have made good around the edges.

## DOORS

In old cottages you can expect to find almost any type or period of door, depending, as with windows, how many times it was found necessary, owing to wear or rot, to replace them.

In pre-Roman Britain a door was simply a loose slab of wood, or wattle screen, which was leaned in front of the entrance. Later, a more substantial 'hanging door' was introduced, harr-hinged, that is suspended by projecting wooden pegs in the top and bottom of the stile which fitted into sockets in the lintel and threshold. It should be pointed out that these hinged doors were no startling innovation, for the same principle was employed by the Ancient Egyptians and Etruscans, who chiselled harr-hung doors out of solid stone for their tomb entrances.

Iron hinges followed, a development of the hook and band which can still be seen on some old field gates. In the early Tudor period the heavy iron strap hinge and pin came into general use. Our modern butt-hinges are descended from the ornate twin cock's-head hinges which varied in design until they reached the more or less standard elongated H pattern, popular during the eighteenth century, particularly on inside doors and cupboards.

The most primitive door fastening still in use is the loose bar, or sneck, which is placed across the inside of the door and secured in slots at both ends. This method of fastening a door was the accepted mode of the majority of smaller houses well into the seventeenth century. Its big disadvantage of course, was that it was only possible to lock or unlock the door from within. In order to overcome this it was found that by drilling a hole through the stile of the door and inserting a string, attached on the inside to one end of the bar, it was possible to lift it out of the slot and allow the door to be opened from the outside. To secure the door against being opened by pulling on the bobbin end of the string—in Shropshire called a 'clicket' and in other parts a 'sneck-band' or 'snecket'—it was just a matter of wedging a piece of wood, called a 'snib' in the staple so that the bar could not be raised from the outside. The bar was later shortened and hinged at one end to the door itself, with the other end projecting just far enough to fit into a wooden catch on the door jamb. This was known as a sneck-latch and was the original of our present-day thumb-latches.

Door handles with box-lock and spring catch followed as a natural course, along with gracefully fashioned knockers which, during the eighteenth century, became highly developed. It was at this time that brass came into its own for these purposes in the better houses.

As to the doors themselves, apart from more skilful joinery and better finish, they have changed very little in principle since the ledge-and-batten doors of the Middle Ages, when single boards were butted together and secured by nailing or pegging horizontal pieces across the back. The tongue-and-groove panel door is but an extension of the batten and ledge.

Another type of early door still in use today, which was popular during the seventeenth century, is the hatch door, or 'heck-door' as it was once called. Such horizontally divided doors had many advantages. For instance, in houses without windows the upper 'heck' could be opened to let in light; or the lower half to permit the going in and out of domestic animals, such as pigs and hens, which shared the house with their owners.

On old outside studded doors of the Tudor period the vertical face boards very often have strips of chamfered oak filleting running down between the edges to cover the unevenness of the joints as well as to keep out the rain. You may be fortunate enough to acquire a Tudor cottage with a genuine Tudor, studded front door. It is not in the cards perhaps, but it could well happen. It happened to us, or rather we made it happen. We bought a late sixteenth-century cottage in the West Country a few years ago. It was part of a big old family estate that was put up for auction. The cottage we were after, and subsequently managed to get, was in a shocking state of repair: the roof, once thatched, was almost bald and had started to fall in; the cob walls were beginning to disintegrate for lack of care, and the front door, rickety and half rotten, was hanging by one rusty Victorian hinge. However, we accepted the challenge to restore it as nearly as possible to the way it was almost 400 years ago.

While we were clearing away some of the debris and wondering just where to start and how to go about it, a representative from a building firm in a nearby town stopped by to look it over and offer a few words of professional wisdom. 'Only one thing to do with a pile of old rubble like this,' he said, proffering his business card. 'Send for a bulldozer and clear it away. It's got no right to be still standing. Now we could put you up a smart little bungalow on this site.' We chose to ignore both his card and his advice and carried on, determined to go through with our original idea.

The front door presented us with quite a problem. The existing door was far too small and was in such a terrible condition that it would have disgraced a chicken coop. Then one day a friend told us of an interesting old door he had seen fitted to a barn on a dilapidated farm a few miles away. We drove out and looked at it. Half hidden by undergrowth and ivy, which had entwined itself around the rusty iron pear-drop handle, was a genuine studded sixteenth-century door, set in a massive oak frame on which were carved Tudor-rose emblems. The door in its present position seemed to serve no useful purpose, for it obviously had not been opened for many years. As far as we could see its condition was amazingly sound, and we decided then and there that somehow we must have it. Luckily, the farmer showed little concern for the door itself—to him it was just part of the wall—and when we made an offer of £10 he jumped at it, making the one condition that we would be responsible for taking it out and making good the wall afterwards.

The next evening our builder's men brought the door and heavy frame, or durns, back to our cottage in their truck, having carefully removed it from the wall and filled the gap with cemented-in stones to the farmer's satisfaction. How it ever came to be installed into the side of a crude stone barn in the first place we were never able to find out. Possibly it was taken from a derelict old yeoman's house nearby and used just as an in-filling.

It took almost a week's hard work to restore the old door to something like its original appearance. The stout heart-of-oak planking, backed with wide elm boards, was, considering the long years of neglect, very little affected by worm or rot; and the heavy wrought-iron strap hinges and stud heads had suffered only little surface erosion from exposure.

Continuous wire-brushing along the grain, an arm-aching procedure, and seeming interminable sanding, finally removed the last vestiges of oxblood and nut-oil paint with which it had at one time been coated. It then remained to treat the whole door and frame with several coats of light machine oil and insecticide, with just a little dark stain added to give the colour an even finish. The iron parts we touched up with a weak solution of matt black oil paint to enhance the contrast; too much artificial colouring on aged wood or iron mars the natural appearance.

The huge door frame, with its low-arched Tudor lintel and stop-chamfered edges, weighed several hundred-weight, and in order to excavate enough room to house it was necessary to shore up an area of 10 ft or so around the doorway. But all this was accomplished successfully and the wall was rebuilt tight to the durnposts and lintel.

Today that 'pile of old rubble that had no right to be still standing' is the object of great fascination to people who come from far out-of-the-way places to admire it and take pictures to remind them of the charm of a quaint Elizabethan cottage, with its roof of thatch and studded oak door, so in harmony with its rural surroundings that it might have grown out of the ground. It was hard, tedious work, but we looked on it more as a labour of love. And it gave us the deep satisfaction of knowing we had preserved a tiny bit of old England for posterity. A solid home, and likely to remain so long after many a modern plastic-and-plasterboard creation had passed the point of no return.

If you decide your front door is too small, which might well be the case, or if you feel it does not do justice to the cottage, then before you have your carpenter make one up or order one from a supplier, enquire around of demolition firms. They may be able to provide you with one from their yard, or from an old

*Page 87* (*left*) Exposed massive cross-panelled black oak beamed ceiling with oak and elm screen, partitioning off through-way. Originally a thirteenth-century chantry. Until recent restoration the building was in use by a farmer to house pigs and sheep with battery poultry on the upper floor; (*right*) exposed beams, with corbell end-supports, of fifteenth-century yeoman's house

*Page 88* (*left*) Cruck members and purlins framing roof of old hallhouse showing corridor leading to bedrooms on built-in upper floor; (*right*) exposed curved cruck and purlin of early fourteenth-century farmhouse. The staircase, recently added to give access to built-in upper floor, was designed to conform to the original hallhouse frame

cottage they may be in the process of demolishing. Doors are to be found—if you have the patience to search. Above all, keep away from cheaply faked or far out-of-period replacements.

CHAPTER 6

# The Interior

We have so far concentrated only on the outside of the cottage. Let us go in and look round the inside to find out what needs to be done to make it comfortably livable, attractive, and in character.

The average country cottage as mentioned earlier, consists of two rooms downstairs, with a small kitchen off, in many cases fashioned from a lean-to annexe. There is usually a narrow, sometimes winding staircase which breaks on to a small landing a couple of stairs from the bedroom level, two low-ceilinged bedrooms, and a small box-room. The latter may have been converted into a more or less makeshift bathroom. It must be realised that there is no set pattern for the geography or size of any of the rooms, as it is possible that alterations and additions have been made from time to time since the cottage was built. Furthermore, the architecture of old rural cottages varied to some degree according to the locality and materials at hand. But let us assume that if and when alterations were made by previous owners they were done at one of the periods in the history of house development after the Renaissance most likely to have affected the rural cottage: (1) during the latter part of the eighteenth century; (2) the Victorian era; (3) since World War I. Briefly, these are the reasons for choosing these particular times.

(1) The spurt of commercial growth which took place towards the end of the eighteenth century infused into the social order a renewed stability in place of the old feudal system. Wealth was concentrated into fewer hands, and those who possessed it spent it lavishly, building immense country homes of unparalleled grandeur, incorporating the newest techniques and architectural designs, including the early use of structural iron, all executed with superb craftsmanship.

With the digging of the canal system and the subsequent development of steam power in factories and mills it became possible to transport coal, iron, and goods over a very large part of the country. While the poorer classes had little or no chance of improving their housing conditions, they were at least able to take advantage of some innovations and copy them, crude perhaps though they were. Consequently, items such as windows, doors, skirting boards, flooring, and staircases were available from local timber mills, machine sawn and mortised from imported soft woods. Joinery work improved accordingly. Also styles began to change at this time. Window frames now, instead of being set flush with the outside wall as in earlier periods, were installed with protruding lintels and sills. Doors were mill-cut and dressed, with tongue-and-groove seatings, and the employment of ovolo wood moulding became commonplace.

(2) With the advent of the railway at the beginning of the Victorian period even wider distribution of materials was possible. Factory-made bricks and slate roofs were becoming the accepted materials for house construction, especially since the new Portland cement had just come into popular use. Steam-powered timber mills outdid themselves with fancy rococo designs of machine-cut fascia boards to decorate soffits, outside eaves, and balconies, and foundries produced ornate cast-iron grills and railings to match. Mass production was getting under way. Factories, now equipped with the new gas-lighting, worked long hours to cope with the demand. Soon another time and labour saving material was being manufactured and exploited to the full—corrugated sheet galvanised iron. Machines were now turning out many times the quantity of materials in far less time than hitherto, though the old hand craftsmanship suffered as a consequence.

(3) The first quarter of the present century saw electric lighting replace the old gaslight. More modern ideas began to be incorporated into house building. Fireplaces were more or less standardised with coal-saving grates and low boxed-tile surrounds, doing away with the mantelpiece. Bathrooms, heretofore considered a non-essential requisite to the average house or cottage, were now a 'must' for all new dwellings. The old-fashioned scullery, almost invariably always cold and damp, was found to be an unnecessary appendage and expense for the builder, and was done away with.

By carefully going over an old cottage you can

usually detect the various alterations and establish, to a close approximation, their period. However, as we have found, there are those old houses and cottages that have been altered and repaired so many times over the years that little of the original is discernible; in fact they can almost be compared with King Alfred's axe, as explained by a museum guide—'It's the original article, only it's had two new heads and three new handles.'

## FLOORS

The earliest floors, of course, were the natural ground on which the building was erected. It is perhaps hard to realise that, apart from a comparatively brief period when the Romans used tiles to floor their villas during their occupation of Britain, common earth was the accepted material in homes, large or small, for well over 1,000 years. In fact, variations of earthen floors continued to be used well into the eighteenth century.

Elementary as an earthen or mud floor seems to us, the technique of preparation and laying rose to a technical excellence. One Henry Best, in the seventeenth century, wrote a detailed description of his method of laying mud floors. The earth was to be dug and raked until the moulds were 'indifferent small'. The water was to be brought in 'seas' and in 'great tubs or hoggsheads or sleddes'. The earth was then to be watered until it was a 'soft puddle'. It was then allowed to lie a fortnight until the water had settled and the material had begun to grow hard again. Then the floor was to be 'melled' and beaten down with wooden paddles to a smooth finish. The generally used mud floors were porous, and so absorbed any wet matter, particularly that of nitrous content; for no one was fastidious about sanitation. In order to obtain nitre, required in the manufacture of gunpowder, an official known as the 'saltpetre man' had power by 'common seisin' to enter any building with his diggers and take up the floors or 'throw down mud walls' in his search. Although there was a provision that such damage should be paid for by the chamberlain for 'makinge the florthe ageyne of newe', it was not always enforced, and much inconvenience was suffered by householders, who came to hate these 'high-handed destroyers'. There is a record of a 'saltpetre man' even digging up the floor of a church in 1624. Not until the time of the Commonwealth was the abuse of saltpetre-digging finally swept away.

Mud floors were very dusty and difficult to keep clean. In dry weather they were often strewn with rushes and damp plants. Some of the better-class houses had mud for the floor mixed with a proportion of bullock's blood, fine clay, and bone chips, which dried hard and gave it the appearance of black marble when polished. This composition, without the bone chips, was also recommended as a plaster for walls.

An improvement took place when lime was incorporated in the floor material; although in some districts, where lime was plentiful, floors containing lime have been made since the Norman period, it was only put into ordinary buildings at the time of the Napoleonic Wars, and the use of sand and lime as a flooring for cottages was new to most areas until well into the eighteenth century. In 1807 a certain Thomas Rudge of Gloucestershire mentions the 'new fashion' floor composed of lime and ashes or sand, laid in a moist state to the thickness of 4–5 in and rammed with a heavy wooden slab until it acquired hardness and a smooth surface. In some parts of the Midlands today such lime-ash floors are still in existence in the upper storeys of seventeenth-century houses which belonged to the lesser gentry.

The employment of timber as a flooring material, common in houses today, can be traced back to Romano–British times, when rough-chipped oak boards were laid tight together and plastered smooth with a covering of mud. The makers of these floors had no understanding of the need for an air circulation. Flag-paving was also used at this time, chiefly for stables and yards.

The earliest existing examples of board and joist floors are in the upper storeys of the keeps of some Anglo–Norman castles. They also used cross-supports, beams laid on their broad sides at right-angles to the joists, from which the more advanced 'framed floor' later developed. The cross-beams were known as sommers or summers, and the main beam that carried the chimney was the bressummer or breast-saddle. Odd as it may seem, it was not until board and joist floors had been found convenient on the upper floors that they were used on the ground-floor level, taking the place of the large stone slabs which in many houses had superseded the beaten-mud floor.

At first boards and joists of upper floors were not concealed, and showed in the room below. There were exceptions to this type of construction, especially in some cottages and farmhouses in the West Country, where rough-hewn joists, often with much of the bark left on, were laid a few inches apart and packed in between with a mixture of clay and chopped straw on interlacing hazel sticks. It was then plastered smooth.

During the sixteenth century ceilings of plaster on laths were introduced from the Continent and became popular in the more expensive homes. Gradually the idea of plastering ceilings caught on and came into more general use in houses of lesser quality. The early laths were hand-riven and rough in appearance. Even so, they were stronger than present day sawn laths of similar material, for with riven laths the grain was not cut through. Plastered ceilings, or underdrawing as it was called, were often filled with chaff or straw to act as insulation.

It is an interesting point that until the thirteenth century ordinary buildings were only single-storey constructions, and even in London houses seldom if ever exceeded two storeys, including the basement. It took another 200 years for upper flooring to be generally acceptable to householders.

The type of flooring of your upstairs rooms will depend mainly, of course, on when it was put in and the material easiest to procure at that time. It could well be floored with 12–14 in elm boards which, perhaps under favourable circumstances, were laid when the cottage was built, and are still serviceable. Or if the original boards have been renewed, then the present flooring must be of a later date, perhaps mill-cut from either local timber or imported soft wood.

Check the condition of the boards carefully. Look for evidence of woodworm and rot, particularly under windows and in corners. Note if the worm burrowings have been freshly made; you can tell by lightly blowing on them to see if traces of light coloured wood dust appear. If not, it is possible that any infestation has been arrested. Nevertheless, every precaution should be taken.

To test the firmness of the joists, stand in the middle of the room and bounce up and down a few times. If the floor responds with a soft, sluggish reaction then it may be that you will have to replace a few joists. Take a wrecking bar and lever up a couple of floorboards. You may have to remove the skirting first. Have a good look at the condition of the joists below, and the underneath sides of the boards. All being sound, you can safely nail the boards back. If there is any doubt and the general condition is ominously suspect, then there is nothing for it but to take up all the boards, first marking them for replacement, either by numbering them separately or by chalking a couple of lines diagonally across the floor for matching purpose later.

Remove the nails from the joists and boards and stack the boards neatly at one end of the room. Now inspect the joists. Take out any that are hopelessly unserviceable and fit replacements. It may be simply a matter of cutting away odd spots of worm infestation or rot. After you have gone over every timber and given it a good wire-brushing, check the seating of the joists at either end to see that they are all firm and of a common level for seating the boards evenly. Now go over everything with a couple of coats of dual-purpose insecticide and rot fluid, and don't be sparing. It may even be necessary to apply a fungicidal fluid on any area subject to traces of rot.

Replace the boards in their right order, butting them tightly together, after having treated the undersides and edges with insecticide. Nail the boards back in position, using 3-in flooring brads for the purpose, and sweep the floor well. Mix a little dark stain with insecticide and thoroughly treat the floor. If you have had to be remove any of the skirting boards, coat the backs of them with insecticide fluid before resetting.

The downstairs floors are more likely to be of a solid composition such as flagstones, quarry tiles, or lime and ash aggregate, for the reasons already given. Of course, it is possible they may have been boarded over at a more recent date. But assuming they are solid and are in good shape, with no traces of damp, then there is no reason they should not be left as they are with just the surface waxed and polished. However, if the general condition is very bad and does not warrant preserving, then it might be better to take up the floor completely and lay a new one.

To do this pry up the stones or tiles or, in the case of a composition material, break it with a heavy hammer, and clear out a foundation to the walls. The depth to excavate will depend a lot on the original base material and what damp conditions you may find. If the foundations are dry, allow for a hardcore base of 4–5 in rough concrete (one part cement, four parts small stones or fine rubble, two parts sand) to bring the level up to an inch below the surface thickness of whatever finish you want. Old quarry tiles are 1 in thick, almost twice the thickness of ordinary 6 in red floor tiles.

Before you top up the floor with screeding first lay a damp-proof membrane, completely covering the base layer of rough concrete, in the form of either polythene sheeting or a heavy coat of bitumastic paint. Now mix up a batch of strong cement (one part cement, three parts sharp sand) and spread it on top of the rough concrete to raise the level the extra inch, still remembering to allow for the surface thickness. Make sure your level is true by tamping the wet cement with a length of straight batten, preferably with the assistance of someone holding one end, and trowel to a smooth finish. Check the surface with your spirit level. It is a good idea to run a wide strip of damp-course felt round the base of the inside walls before concreting up to it, as an extra measure of protection.

As to the type of floor you want, now is the time for decision. You may want a hard, smooth-trowelled cement surface, coloured or natural, on which to throw woven grass mats or a cord rug. Cement-colouring powder is available in several shades and is mixed in with the screeding aggregate to the proportion recommended by the manufacturers. When polished this makes both a serviceable and attractive floor. Or you may prefer a type of Marley or similar tile surface, in which case the cement screeding must be brought up to within a quarter of an inch of the finished level. There is also on the market an asphalt composition for hard flooring, which comes in a variety of colours and polishes to a wax finish. But to lay this particular material requires the technique of a professional. So before you attempt to do it yourself call in your builder for advice.

To lay a board floor set your joists firmly and level, 16 in centre to centre, butting the boards tightly before nailing and seeing that all the wood is treated with insecticide. Also make sure there is sufficient air circulation. If necessary make vents through to the outside walls and fit in airbrick ventilators.

## STAIRS

Access to the upper storey originally was, as often as

not, on the outside. The earliest method of ascending and descending was by means of an upright pole with pegs driven through it and projecting on each side for the hands and feet. In the early fifteenth century the pole was superseded by a ladder or 'staeg'. This method of approach had the added advantage of being removable from either above or below as a means of security. The first staircases were cut from solid baulks of oak in straight flights. Later each step was cut out in a separate block and set in position. There soon followed a combination of these two types, circular newel steps, in which the newel was a straight upright tree with the steps framed into it. This form of staircase became common during the Renaissance period.

The modern staircase, with separate treads, risers, and close strings, found today in the more humble old dwellings, is a fairly sure indication of a date around the time of Queen Anne. Staircases with cut strings, that is, the side pieces sawn to fit each tread and riser, came in a century later. Where in smaller houses and cottages the problem arose of finding enough room for both staircase and fireplace on the same wall it was solved by going back to the traditional newel stair, turning through two right-angles between the ground floor and the first floor. An alternative method was the fashioning of a dog-leg staircase, whereby it angled off in two directions from a small landing near the top.

About the time of Elizabeth the axial chimney-stack was introduced, that is, fireplaces built back to back in the middle of the house with a common stack or common upper flue. Some yeomen's houses and more substantial cottages followed the lead of the wealthier properties and had central fireplaces built in, at the same time incorporating a stone or brick newel staircase which also acted as an additional support.

It should be borne in mind that almost all new ideas in house building spread from the eastern counties, which in turn adopted their innovations from the Continent. Consequently, the more sequestered areas of England were the last to learn about and accept change; and a great deal of pre-nineteenth century architecture and general house development, particularly in the south-western corner, was often several decades behind the more populated districts to the east and north.

When you are planning alterations or improvements to the inside of your cottage consider carefully the positioning of the staircase. It is possible that geographically the staircase could be situated elsewhere and be more convenient as well as perhaps taking up less room. Or again, perhaps it could be boxed in on both sides; or, to go to the other extreme, it might be completely exposed, not only to give the effect of enlarging the area, but to throw a lot more light all round. Naturally, any modifications of this sort are dependent on the shape and juxtaposition of the staircase in relation to its immediate setting. If it is a matter of siting a new staircase altogether, study all the implications involved and make every allowance for turns and headroom.

If you feel capable of doing most of the work yourself, hesitate before you attempt the carpentry. Almost any joinery shop will be glad to send a man out to measure for a new staircase to your specifications and make one up ready for installation. This will probably prove a lot less expensive than employing a master carpenter to build one by hand.

## INTERIOR WALLS

If the interior dividing walls of your cottage are original and are 12–14 in thick, then the chances are that they are of smooth-rendered cob, coated with several applications of whitewash. If, on the other hand, again provided they are original and pre-date the common use of bricks, they are only 5–6 in thick, you may well have a wattle and daub partition. Wattlework, the weaving of twigs in and out of spaced upright stakes, was used by the Romans, who daubed on a pug aggregate to make solid screens. Wattle and daub was not an original Roman idea, for traces of mud-covered basketwork have been found in the remains of prehistoric dwellings.

From the Middle Ages until, in some country districts, the nineteenth century, the making of wattle and daub walls altered little. The rods and twigs were usually of ash and hazel. Straw was mixed with a sloppy mud or clay and applied to the wattle by two men, one on each side, who kept throwing it until they got the thickness they wanted. Each face of the wall was then combed to a smooth finish. Where there were horizontal timbers or studs set vertically in the wall their surfaces were sometimes left exposed, although in most cases they were plastered over.

The widespread use of wattle and daub in England is shown by its varied names in the old dialects. In Lincolnshire wattle and daub is called 'stud and mud'; in Lancashire 'rad and dab'; in Devon it was known as 'staking and freething'; in Kent 'raddis-work'; and in the northern part of the country and in Scotland wattle and daub was called 'stake and rice', the rice being brushwood on which the mud was applied.

Later some inside walls were built of small stones or rubble, bonded with a mixture of clay and cow dung and plastered smooth. The use of this particular mortar came to be more general and was applied to the inside of chimneys and, especially in Norfolk and East Anglia, to bond early exterior brick nogging, or for infilling between timber framework. This work was known as 'pargetting'.

To repair damage or flaking on interior cob walls, or where damp may have penetrated and bulged out the surface, dig away all loose material until you are satisfied the edges are firm, and leave a ragged surface so that the filler to be applied will make a good bond. Don't be alarmed if you have to gouge out several inches. The surrounding cob will hold tight without any fear of collapse.

Now mix a batch of fairly weak Walcrete (about five parts sand to one part Walcrete) and trowel well into the cavity to within half an inch of the surface. When it has dried apply a slow-setting hemi-hydrate plaster, such as 'Thistle', which is quite cheap to buy and is simply mixed with water into a paste, and again trowel flush with the surface of the wall. Simulate as far as possible the original contour and texture. It is not advisable to apply gypsum plaster directly on to the cob as the materials are alien and liable not to 'marry up'.

The sort of skirting you may expect to find around the base of the walls, if there is a skirting, depends largely on damp conditions, when it was put in, and the kind of floor. If the floor is stone-flagged then the skirting will probably be of tile, which is more in keeping with the floor and also easier to fit. A wooden skirting would be more suitable to a board floor, though they could well be mixed. By examining skirting boards it is fairly easy to get an idea of their period. Many old cottages skirted the base of their inside walls during the eighteenth century when milled boards became extremely popular and cheap; the majority of skirting boards of this period are quite deep, 7–10 in, with fluted or convex mouldings on the upper edge. A further confirmation of date can be made by carefully examining the nails that hold the skirting to the wall plugs. Until machine-cut nails were produced in quantity for general usage around the turn of the last century, they were either made entirely by hand or stamped out by a manually operated press. Modern cut iron nails are clean-edged and evenly flat with a single-flange head of the same thickness, but earlier nails either have roughly rounded heads or double flanges. Wrought-iron studs and nails of earlier times were individually hammered out on the anvil, their heads being shaped while still hot.

CHAPTER 7

# Wood Rot

---

With any wood in direct contact with damp walls or floors, particularly if there is no air circulation, there is always the possibility, if not probability, of deterioration through wet or dry rot. Of the two, wet rot is much easier to deal with. It is always better to arrest the cause before simply replacing new boards, or it will be just a matter of time before you have the same trouble all over again.

## WET ROT

The incidence of wet-rot fungus can be attributed directly to continued exposure to wet or damp, and if not treated can spread to wood in contact with it. It can, if not totally arrested, develop into dry rot, which is even more serious. Test any suspect timber with your knife to determine the extent of sub-surface breakdown and then remove or cut out all affected parts. Be sure to remove every vestige of dust and debris in the area affected, and burn it with the rest.

If wooden floors have wet rot in the skirting, take up any affected boards and examine the floor joists; these, if there is adequate ventilation, should be dry and sound. Under ordinary circumstances there should be little or no trace of wood-boring insects where there is no easy access to outside air, such as wide gaps between the floorboards. However, if worm infestation has affected the joists and underside of the boards it is worth prising them up and treating all the suspect wood with insecticide before renewing or replacing.

Having cut out all timber with, or suspected to have, wet rot, select thoroughly dry and well-seasoned replacements and treat with a couple of coats of fungicidal fluid. Also make sure that all the old wood with which the replacements come in contact is equally generously brushed with the protective fluid. Before you make good any timber replacements first try to establish the cause of the wet rot, so as to remedy it.

If wet is seeping through the wall at the base it is a matter of digging down and cleaning out to the foundation. This may be difficult, as the rough stone footings or sills of some old cob cottages may have settled a foot or so into the ground, so that the bottom of the cob is below the soil level outside. Dig a V-shaped trench against the wall on the inside as far below the ground-floor level as seems necessary, to the stone ground-sill if possible, with the top of the trench about 5–6 in wide. Fill the trench with a Walcrete mix, working it well against the cob. Fill that part of the wall above the floor level to within half an inch of the surface and trowel flush later with an anhydrous waterproof plaster. The cause of the interior-wall damp in the first place can, in all probability, be traced to lack of proper attention to the outside base. But now that it has a good skirting and apron no further trouble should be experienced once the existing damp has had a chance to dry out.

While on the subject of wet rot inside the cottage, check all the other woodwork—window ledges, window seats, frames, inside cupboards, and under the stairs, and proceed as described for floors and skirting boards.

## DRY ROT

In the treatment of timber deterioration we have so far dealt only with wet rot and woodworm. But by far the most serious problem to face in wood breakdown is dry rot, or 'merulius lacrymans', the surveyor's nightmare. Once established it will spread rapidly and extensively and can destroy all timbers attacked. Like wet rot, true dry rot is a fungus which thrives in conditions of warm mild damp and inadequate ventilation.

We were about to buy a charming little Queen Anne house from an elderly spinster a few years ago—until we made our own survey. We had become a little suspicious of a faint sickly-sweet smell which emanated from behind the high panelling in the lounge and determined to explore deeper. We learned that there was a cellar to the house, the door and outside ventilators of which the owner had sealed up because of her hatred of draughts. However, we managed to obtain access to the cellar, and with a torch made a thorough inspection. The result was startling. The walls from floor to ceiling, overhead joists and boards, in fact everything, were literally covered with myriad mushroom-like growths glistening in their own moisture. It was a perfect breeding ground for dry-rot spores. Needless to say, our interest in the house disappeared at once. We learned some months later that it had been sold and the new owners were obliged to spend several hundred pounds in eradicating the infestation and for repairs. The fungal attack had actually spread up the walls, behind the panelling and wallpaper, as far as the upstairs bedrooms. This particular example of how the ravages of dry rot can affect the unwary house-hunter is perhaps exceptional. Nevertheless, it is wise to take every precaution—and take it early.

Whereas wet rot is easier to distinguish by its usually soft 'worn-away' and damp appearance, most often dark brown in colour and, as a rule, found in places subject to constant wet, dry rot is often more difficult to detect. The places to look are in unventilated cupboards and areas such as skirting boards, panelling, and window seats. Don't be misled by smooth outside surfaces of glossy paint or varnish. The fungus likes to work from the back where it is moist and dark and confined. Roofs are seldom attacked by dry rot, even though they may be exposed to wet and are dark, as there is almost always a sufficient movement of air to prevent its taking hold.

Dry rot can be determined by its peculiar breakdown of the wood, which looks desiccated and shrunken and shows transverse cuboidal cracks. It is pale brown, often showing traces of tiny white spores, and brittle, so that if crushed between the fingers it will disintegrate to a fine powder.

The treatment is first to cut out all the wood showing signs of its presence, and also any wood within 2 or 3 ft of the nearest decay, even though it may appear to be sound. Then carefully examine the area well beyond the affected or suspected parts. It is imperative that you make a thorough job of the eradication. Hack away all surrounding cob, earth and plaster, including brick and stonework if it is contiguous. When this is done go over the whole area with a stiff wire brush, remove every particle from the house and burn it immediately. So likely is dry rot to re-establish itself if any spores are left that as an added precaution it is advisable, where possible, to use a blow-torch on and around offending cob or stone areas.

Treat everything, wood, stone, cob and earth, to a good soaking with fungicidal fluid, making sure to penetrate every crack and surface. Now re-render with Walcrete, as with the treatment of wet rot, and apply a thin coat of zinc oxychloride plaster over the Walcrete before you plaster flush with 'Thistle' or other comparable hemi-hydrate finish. This will give you added protection against any likelihood of recurrence of the dry rot fungus.

Before you fix any new wood to replace the parts you have cut away make certain you soak it thoroughly with fungicidal fluid.

CHAPTER 8

# Fireplaces

---

There is nothing so cosy and welcoming in a small cottage living room or lounge as a fireplace with good old-fashioned character, especially if it is set off with exposed beams and ceiling joists. We have mentioned briefly earlier the general introductory period of the central fireplace when brickwork, particularly in the eastern counties, became popular in the building of flues and chimney-stacks. But in the more remote parts of the country stone and cob continued to be the accepted materials, both because they had always served the purpose well and because they were a lot handier and cheaper to come by than kiln-baked bricks.

Generally speaking, fireplaces with built-in flues and chimney-stacks, primitive advancements though they would seem to be today, are comparatively recent. Until the sixteenth century the vast majority of

dormer windows, a through passage and cross-wing consisting of a buttery and service rooms, all either completely chambered over or at least with a solar or loft. Often a separate wing, or even a separate building, was built on for the kitchen, which was considered of secondary importance in many houses to the more necessary proximity to the main structure of the shippen or barn.

Most smaller sixteenth-century houses and farmhouses of any substance had through-ways if not cross-wings, with a fireplace in the hall backing on to the through-way wall and another in the parlour on the outside wall. The parlour was often later converted into the kitchen, and full advantage was taken of the huge fireplace by incorporating a bake oven and smoke chamber.

There seems to have been no set pattern for the positioning of fireplaces. Like the house plan in general, laid out between owner and builder to a near-enough set of specifications, the fireplaces were situated more for convenience than to comply with form or function. There are exceptions, of course. Some old Elizabethan inns and two-storeyed jettied houses, as well as cottages, were carefully planned in advance, and the chimney structure shows that much thought must have been put into the architectural design both for utility and appearance.

## EXCAVATING AND ENLARGING

But back to our cottage. We will assume for our purpose that there is a fireplace in each of the gable end walls, their flues linking up with small fireplaces in the bedrooms immediately above; and, if there is a third room on the ground floor, another back-to-back fireplace running up to an axial stack in the roof.

Now a compact Victorian or even more modern tiled fireplace can look quite attractive in a cottage dining-room, and the kitchen flue is, naturally, strictly utilitarian for the range. But the fireplace in the lounge living-room, especially if the joists and beams are exposed, should be the focal point of old-fashioned charm and comfort. It is probable that your particular lounge fireplace is a small nondescript affair with a tiled surround. We want a large open hearth where, if necessary, we can toss on logs to blaze freely instead of delicately feeding the tiny grate with nobs of coal from a pair of brass tongs. So a major operation is called for. There is really no need to be afraid of a job like this, providing you take it step by step and employ a measure of good common sense.

First assess the depth of the original fireplace, making allowances for the thickness of the wall. Remember that if the fireplace was built with the cottage it would have been a simple cavity of fairly generous proportions, probably 4–6 ft across and anything up to 3 ft or so deep. Probe through the plaster directly above the fireplace, 4–5 ft from the floor, to find if the original clavel, or fireplace lintel, still remains embedded in the wall. If it is still there you are in luck, especially if its condition is still comparatively sound. If not, one will have to be built in. But we'll come to that later.

With a lump hammer, a small wrecking bar, and a bricklayer's steel bolster, take out the existing grate and tile surround and strip off the plaster to expose the brickwork. The fireplace may have been put in or modified within the last twenty or thirty years, in which case it is probably built around an older Victorian-type fireplace underneath. It is usually easy to tell the difference in periods by the brick and mortar work. Immediately over the fireplace opening there is in all likelihood a flat iron lintel bar, put in as an initial support for the bricks, and perhaps a register plate with an iron flap to shut off the chimney.

If there is a sound clavel beam above to support the weight you can tap away the infilling beneath, beginning in the middle and working outwards in the shape of an inverted V until you have exposed the whole underneath side of the beam. Make certain there is sufficient hold on the piers at either end. (These were generally built of stone.) If there is no clavel beam in the wall, or if the existing one is in too bad a condition to expose, then, before you excavate any further, it is essential you replace it or fit one in. This is not such a difficult operation as it may first appear, so long as you go about it the right way.

Should it be just a matter of replacing the old beam, first prepare a suitable replacement of as nearly the same proportions as the original. If you can run across an old timber which can be cut to size so much the better; but a good alternative is a length of old railway sleeper, well sanded, treated with insecticide-stain, and chamfered on the inside towards the bottom so that the edge is about 3 in wide. With a replacement clavel beam it is just as well, as a safety measure, to line the inside face and inside edge of the bottom with asbestos. Cut a piece of sheet asbestos to size, using an old saw for the purpose, drill the screw-holes, and fit neatly so that it is unnoticeable from the front.

Carefully chip or dig out the old beam from the cob or stone, being careful not to hammer too heavily to disturb the wall above. Fit the replacement in position, set on its edge and brought forward so that it is proud of the wall about $\frac{1}{2}$ in. Seal it all round with Wallcrete, pressing it well home back and front to secure a firm seating. Once the Walcrete has set hard you can clean out the fireplace below the clavel to the original back and side walls. They will be either of stone or plastered cob and probably still show evidence of soot marks. The original hearth floor may be of stone, small quarry tiles, or simple earth.

Now let us consider the prospect if no clavel beam is *in situ*, or, in fact, if one never existed—as is sometimes the case with small fireplace openings where they relied solely on low, keyed brick or stone lintel arches. So long as the chimney breast is wide and deep enough a beam can be fitted to enable the fireplace to be enlarged. Have your beam prepared in advance, with the inside face and bottom edge asbestos-covered. Mark out on the wall over the fireplace the position in which it is to be installed. A good positioning would be 3 ft 6 in to the hearth from the bottom of the beam, and 4 ft wide. Allow 6 in of overhang on each side to seat on the piers.

Cut out a space in the breast wall with your lump hammer and bolster for the beam to fit, starting in the centre and working outwards as before. You will need patience for this job, for it must be done with great care, keeping an eye on the wall above to make sure it stays tight. Under ordinary circumstances you should experience no difficulty in making a clean cavity; well-laid cob will bond like reinforced concrete. If, however, the wall contains a fair amount of stones or is perhaps of more or less solid stone, then it is worth taking an extra precaution. First have ready a couple of flat iron bars, approximately 2 × ½ in, and a few inches longer than the clavel cavity. Now cut out a slot in the wall an inch or so above the seating mark along the top edge. Insert the iron bars with about an inch separating them and pack the ends snug with rubble or wooden plugs. Be sure to see that the ends are supported in the wall at either end beyond the cavity. The iron bars, acting as a dummy lintel, will take the weight of any stones or rubble that might come loose immediately above them. You can now cut out the remainder of the cavity below.

When the excavation is finished, gently fit the clavel beam in position and seal in all round, including the iron bars, with Walcrete. Press the Walcrete well home into the crevices, back as well as front of the clavel, allowing for the finishing coat of plaster which will be applied later to the face wall.

We now have a fireplace opening 3 ft 6 in high, 4 ft across, and 2–3 ft deep, running up to a clear vertical chimney shaft. A fire in such a large opening would probably result in a smoke-filled room each time the wind changed direction, even if the draught was satisfactory. Therefore, we must concentrate the up-draught over a smaller area and towards the back, so that it remains constant.

The style of fireplace to build is largely a matter of personal taste. The simplest, of course, and one that is cosy and throws out a good heat, is an iron fire-basket, or a pair of dogs, with a metal hood above. Or again, doing away with the fire-basket or dogs and retaining the metal hood, you can hollow out a saucer shape, about 4 in deep, in the hearth and make an elementary Devon grate. Another alternative is to build a brick or stone hood with corbelled sides. But whichever type

STONE HOOD WITH CORBELLED SUPPORTS

METAL HOOD FIXED TO BACK WALL

you decide upon you must first fit a register plate in the chimney shaft for the flue pipe.

For this you will need half a dozen or so iron spikes about 8 in long (round an old cottage it is usually an easy matter to find pieces of iron suitable for this purpose, such as old strap hinges, worn-out files, Suffolk latch bars, etc), a length of flat iron bar a few inches longer than the width of the inside chimney and roughly 2 in wide, and a piece of heavy gauge asbestos sheeting.

Measure the dimensions of the chimney shaft a foot or so above the top of the clavel beam. Saw the sheet asbestos to these measurements so that it blocks off the chimney completely. Hold it in position and mark on it the middle of the fireplace 5 in from the back wall. With your mark as a centre, cut out with a hacksaw blade, after first drilling a starting hole, a circle of 6 in diameter to take the flue pipe. Now, allowing about 18 in from one end, saw the asbestos in two across the narrow side, checking that your fluehole is at least a few inches inside the larger piece.

Drive the spikes into the walls of the chimney shaft, leaving a couple of inches showing and so spaced as to seat the register plate evenly all round. Fix the length of flat iron bar across the chimney shaft at the same level as the spikes, so that the ends of both pieces of asbestos fit together on top of it. The bar should be bedded into the wall on either side by digging out enough cob or mortar to make a firm seating, later to be made good with Walcrete.

Remove the smaller piece of asbestos, which is to serve as an access trap for sweeping the chimney, and, with the aid of an electric torch, liberally plaster Walcrete all round the top edges of the larger piece, sealing it tight to the wall on three sides. It should be mentioned that the wall should be first scraped free of all traces of soot where it comes into direct contact with the Walcrete, so as to make a key for a good bonding. Now replace the smaller piece of asbestos and trowel Walcrete all round the underneath edge to the walls on three sides, making a chamfered finish. When dry, the trap will tap away easily from the Walcrete if and when it needs to be removed, and will rest snugly back in place again.

We are now ready to fit the hood. Almost any tinsmith or sheetmetal shop will make up a pressed-steel hood from your measurements and a rough sketch of the design you want. To ensure the maximum draught it is advisable to add a separate 6 ft length of 6 in flue pipe to the collar of the hood so that it extends well up into the chimney shaft. The length of flue pipe should be fitted on to the neck of the hood, through the trap, after the hood is secured in position. To secure the hood drill or punch four holes through the back and with spikes or expanding Rawlplugs firmly attach it to the wall, centring it directly beneath the hole cut for the pipe. The bottom of the hood should be not more than 24 in from the bottom of the fire base. Any higher will make the up-draught less effective.

The original hearth, whether of quarry tile, broken brick, lime-ash, or just plain earth, would almost certainly look better if remade. If you are going to use a fire-basket on a flat hearth dig out an area 2 ft 6 in long, 2 ft wide, and 2 in deep in the centre of the hearth against the back wall and, with lime mortar, fit in a base of firebricks or the larger fire lumps, extending them up the back wall a couple of feet where the heat from the fire will be most intense. Remember

*Page 121* (*left*) Through-way of restored sixteenth-century farmhouse from front door; (*right*) Tudor oak plank door studded to horizontal elm boards on inside

*Page 122* (*above*) Excavate
fireplace relined with stone
and fitted with metal hood
(*below*) dining-room enlarg
removing partition wall in s
teenth-century cottage; not
corbelled supports

to allow enough room between the tiles or firebricks to be able to point them flush later with a proprietary brand of plastic fire cement. Lay the rest of the hearth space with small quarry tiles, well bedded with Walcrete, from the firebricks to the edge of the hearth.

Should the back and sides of the fireplace be of rough or crumbly plaster it is worth re-rendering them with Wallcrete, after first scraping away any loose or flaking material, soot, etc, and scratching the surface to make a key for adherence. If you fancy pebbles, small stones, or perhaps quarry tiles, as a fireplace lining, fit them into a base of Walcrete while it is still wet.

You may prefer a brick or stone hood. When made properly these can be not only most attractive, but every bit as efficient as a metal hood. The only difference is that a little more time and patience will be required in the building. You will need three lengths of flat iron bars, one 3 ft and the other two 2 ft 6 in, each 2–3 in wide and ½ in thick. These are to make a base frame. If the hood is to protrude, as in this instance, 24 in from the back wall with a width of 3 ft, then the two trimmers or side pieces should be 6 in longer to allow for their being driven securely into the back wall. Lay the other length across and on top of the trimmers, making a frame, remembering, of course, to set it in the dead centre of the fireplace and 22–24 in above the fire base. Loop half a dozen short lengths of strong thin wire over the frame at equal distances all round. Attach one end of each tight to the frame with the other ends hanging free. These are to be used later.

Now mix up a fairly stiff batch of cement (three parts sand, one part cement, one part lime) and line the top side of the frame all round. While the cement is wet lay a course of small ornamental face bricks, or stones, as the case may be, and smooth plaster the cement all along the tops. Leave to set. When the cement is hard take a piece of stout wire mesh (garden trellis is ideal for the purpose) and cut and bend it in the shape you want your hood to be, whether round or square, allowing an extra inch against the back wall on either side to be spiked into the cob or plaster, and reaching to the register plate at the top and the edge of the iron frame below. Wire the skeleton mesh to the frame at the bottom with your lengths of cemented-in wire, so that the mesh is inside the course of bricks or stones.

You are now ready to lay up with bricks or stones around the mesh skeleton. Be as generous as possible with the cement as you go and plaster it tight to the wall on either side at the back. If you feel that the iron base frame is not sufficiently rigid to take the weight of the freshly cemented hood, then it is advisable to shore up the front corners with scantling until the cement hardens or until the side supports are built.

To fashion the corbelled sides so that they are balanced, first drive a peg into the hearth floor on each side of the fireplace about 4 in from the back wall and directly under the outside edges of the hood. Tie a string to each peg and fasten each length to the underneath side of the hood immediately above it at a point about midway along it. Cement the bricks or stones in an inverted stair pattern up tight to the hood, using the strings as guides. A good rule to observe with this kind of corbel work is to lay only as many courses at a time as will stay firm, and allow them to set before building the next few courses.

In the same brick or stone you can build in a fender

surround, cementing it in place. Such a fender will not only be effective, but will also match the character of the fireplace. And if there is sufficient space inside the fireplace it is not difficult to fashion an inglenook to your own design to add a little extra touch of quaintness.

## BLOCKING UP

There may be a small old-fashioned fireplace in one of the upstairs bedrooms, the flue of which connects with the main chimney shaft of a fireplace below, and it serves no useful or decorative purpose. Because of the room it takes up it might be better done away with altogether. There is not much difficulty involved in blocking up a small fireplace. With the aid of your wrecking bar, lump hammer, and bolster, pry or knock away the existing surround, tile, iron, or other ornamentation, until you are left with just a gaping ragged hole in the wall. Clear away the hearth to floor level.

If the hearth is already on the same level as the floor and is sound and smooth there may not be a case for taking it up to replace it with floor boards, especially if the room is to be wall-to-wall carpeted. If it is found necessary to do so, then it is just a matter of digging out the hearth and matching in boards, laid over and nailed to a couple of lengths of 2 × 4 in joist pieces. When fitting the new wood allow enough air space underneath to keep the boards from coming into direct contact with earth foundations, and be sure also to treat it all with insecticide in advance.

Using a wide-bladed scraper, scrape the wall edges

flush all around the cavity and clear away sufficient space inside for bricking up. With your mortar mix prepared (mortar of four parts soft sand, one part Walcrete), first lay a strip of damp-course felt on the floor as a base on which to set the bricks. See that the felt is well inside the final plaster line. Now brick up the cavity, allowing for the thickness of a mortar rendering over the bricks as well as for the finish plaster coat. It is a good idea to use a length of straight batten when you render and plaster the surface. After trowelling on the mortar rendering shuffle the length of batten, using both hands, across and down the fresh mortar, keeping each end of the batten tight against the solid wall on either side. This will ensure an even face. Be sure to trowel well against the edges all round. Now scrape away a layer of mortar, say a quarter of an inch, so that when the finish plaster is applied it will come flush with the rest of the wall. While the mortar is still wet, score it with your trowel, making a few criss-cross scratches as a key for the plaster.

As soon as the mortar rendering is dry apply the plaster coat. It is better to use a flat steel trowel for this purpose as it enables you to spread the plaster with a sweeping action and keeps it smooth. Repeat the procedure of shuffling the batten across the wet plaster, and just before it 'goes off', or begins to set, sprinkle it lightly with a wet brush, and trowel to an even surface.

There are a couple of points to consider when applying proprietary brand plaster. If the mortar rendering has become very dry before you are ready to plaster, dampen it first by brushing it lightly with water. Secondly, it is well to remember that commercial plaster when mixed can 'go off' or stiffen up with very little notice. So mix it to a fairly sloppy consistency and add a little lime. This will tend to slow down its speed of setting.

It may be that the fireplace you intend to brick up is on the ground-floor level, in which case it will need ventilation so that the locked-in air doesn't become stagnant and damp. Neat brick-sized sliding ventilators are usually kept in stock in hardware stores. It is quite simple to fit one of these in the middle of the second course of bricks as you work up, and cement it in firmly. At the same time attach a short length of upright pipe with a vent head to the back of the grill face.

If, by blocking up a fireplace, the chimney shaft will no longer have any practical use, then the top of the outside stack should be capped and vented at the side away from the prevailing rain-carrying wind.

CHAPTER 9

# Improvements

## CEILINGS AND EXPOSED JOISTS

Ceilings of old cottages, like the ceilings of almost all buildings common to the same period, are split lath and plaster. Horse or cow hair was usually mixed with the plaster as a binder, even into the beginning of this century, and possessed excellent adhering qualities when properly applied by 'bleeding' it well through the closely nailed laths.

Quite often in cases where the wall foundations may have settled unevenly the ceilings sag or show cracks; or perhaps patches of plaster have bulged or fallen off due to constant water dripping. Perhaps the ceiling, by virtue of the laths being fixed to the bottoms of uneven joists, is a bit wavy. In this case there is little you can do about it except strip off everything and re-level.

That is, if you really want to. After all, an uneven ceiling in an old cottage, providing it is firm, is surely part of its charm and character.

Before we deal with repairing cracks or making good loose or fallen plaster, let us, with fingers crossed, see if there is any chance of doing away with the ceiling as such altogether and exposing the joists. In the lounge or living room of a restored old cottage there is nothing more attractive than a ceiling with exposed joists, providing, of course, their condition warrants exposing.

With a small wrecking bar pry off the plaster and laths in one spot over an area of not more than a couple of feet to begin with. Remember, you might be in for a disappointment and have to put it all back. The joists will be, in most circumstances, as they are today, 16 in apart, centre to centre. Be careful when you go to break off the laths that you don't tear off a strip of plaster the whole length. Old hand-riven laths are extraordinarily tough to snap. Examine the joists closely. They should measure approximately 4 in square and be slotted into mortises in the cross-bearing beams, about 6–7 in above the level of the plaster. Dig in with your knife to test their condition. A bit of worm here and there need not cause you too much concern so long as the wood is sound through the bulk and the seatings are firm. If you are satisfied it is worth proceeding further, then extend the area of exploration over the whole ceiling. If, on the other hand, the joists are found to be far below your expectations and not worth the trouble of exposing, nail back the laths and replaster, ready for decoration later.

Let us take the optimistic view by supposing you find the beams and joists in fairly good shape and worthy of dressing in order to show them off. Strip off all the plaster and laths and pull out as many lath nails as you can. All this can be a messy operation, as there is bound to be plenty of dust, cobwebs, and probably bits of straw to contend with. Clean all the wood with a wire brush, and chip or scrape away any soft spots or surface decay. You will find the floor-boards of the upstairs room are laid across the tops of the joists. These boards should be in fairly sound condition and tightly butted together.

An invaluable tool to have with you when dressing old timbers is an electric hand-drill with a coarse sander-disc attachment. If there is no electricity then the same results can be accomplished by using coarse sandpaper over a wood block, only it takes a little longer and needs lots of good old-fashioned elbow-grease. Sand the beams and joists to smooth down all furry surfaces and ragged edges. Don't worry about trying to sand adze-cuts smooth. They are part of the original character.

When this is done apply a good soaking with insecticide. Brush it liberally into every part of the exposed wood, saturating all joints and wormholes. Then go over it all again with a second coat, this time mixing a proportion of dark stain with the insecticide, depending on how dark you want your wood to be. There may be a few particularly bad spots of timber surfaces where you have had to gouge deep to clean out a decayed area, or where a concentrated group of wormholes looked a bit too unsightly to leave. Mix in a bowl enough cellulose wood filler to a stiff paste, then add a little black cement-colouring powder—plain soot will do almost as well. Work into this mix a few drops of dark stain until the colour matches the wood, and fill the wormholes and cracks, using a putty-knife.

When dry give it a light sanding and touch up to an even finish and colour.

A point to look out for when inspecting the timber structure of the ceiling is the possibility of warped beams having pulled their mortises away from the joist tenons. Sometimes a slight bellying on the outer walls will produce the same result. It is essential to re-set the joist ends firmly before you proceed further. To splice short lengths on to the joist ends in order to re-seat them is both unsatisfactory and dangerous. Remember, they carry the weight of the floor above. The best method of establishing a firm support, apart from fitting new joists of the correct length, is to reinforce the beam on the inside with a length of stout timber and wide enough to seat those joists which are out of line. The timber can be trimmed to shape and cut to match the original mortises so that when securely fixed to the beam with iron pins, treated, and stained, it need not look at all unsightly.

Now it is quite possible that having prepared the ceiling beams and joists, the under sides of the floor-boards above the joists will either be too rough to expose or, if left exposed, may tend to make the room too dark. Also, with the ceiling plaster removed there is the matter of sound insulation to consider, especially if the room directly overhead is a bedroom. To remedy one or both of these problems it is not difficult to fashion a white plaster ceiling between the joists.

Measure the total area of exposed ceiling boards and translate it into square feet of plasterboard laths required. Plasterboard laths are standard size—16 × 48 in. Saw and fit the laths, following where possible the contours of the joists to ensure a minimum amount of gaps, and nail them in position, after, of course, you have first treated the original boards with insecticide. Use sherardized or rust-resisting special nails for the purpose or the nail-heads are liable to oxidise and show through the ceiling paint. If you find you still have a few gaps, neatly fill them with cellulose plaster and level off any depressions made by the hammered nails. Give the plaster laths a couple of coats of white emulsion paint. A 2–3-in brush is best for this purpose, which is to make a clean line along the edges of the joists.

## TAKING DOWN AN OLD WALL

One of the shortcomings of many of the smaller old houses and cottages is that the downstairs rooms give one the feeling of being cramped. Where there are two rooms divided by a common wall, the dining-room and parlour (or front room), give some thought to the practicability of knocking them into one long comfortable lounge, especially if you have enlarged the fireplace, which may now appear rather too large for the area it is required to heat.

Examine the wall to find out whether it is simply a partition, or a bearing wall carrying the weight of the floor above it. Cut away a small section in the middle, where the top of the wall meets the ceiling, and expose the joist. If, as is most likely, the wall is built immediately beneath a main joist or summer beam, which appears to be pretty sound, it is safe to proceed further. On the other hand, if the bearing member has rotted beyond repair then it will have to be replaced. But we will tackle that problem later. Should the wall have been put up as a partition only, to one side or the other of the main joist or beam, it is fairly safe to take it away without weakening the general structure.

A point here apropos of wattle and daub, covered in Chapter 7. As a rule most old wattlework buried in screens and partitions is found to be so deteriorated and fragile as literally to fall to pieces when exposed to the air. However, there is just a chance that you may find, if you have a wattle screen, the original stakes and interlacing twigs in a good enough condition to expose and dress (see plate page 140). If you are so fortunate it is well worth the trouble delicately to remove the daub, picking it away a little at a time, and then to brush lightly until the whole surface is clean. Treat it with insecticide, perhaps slightly coloured with an admixture of stain, and polish with liquid wax, using a soft brush.

Now let us concentrate our attention on the dividing wall and assume it is built directly under a main joist or beam. The condition, on first inspection, is good. After having cut away a small section in the middle, start at one end where the beam is either nested in the outside wall or rests on a cruck timber or pier, and expose the joint to ascertain if the seating is secure. Do the same at the other end. Check everything thoroughly. As long as you are satisfied the whole length has a good bearing strength you can carefully take away the rest of the wall, down to an inch or so below the floor level.

However, if you are in any doubt as to the bearing qualities of the end jointings then it is best to reinforce them. There are several ways of doing this, depending on the material of the piers or supports. If the crossbeam is mortised into crucks or vertical wooden studs it might be better to spike heavy iron angle-brackets at either end. If the beam is bedded into a stone or cob wall, or supported by stone or cob piers, then it is a simple matter to build up from the floor a brick extension tight to the underside of the beam. Render the brickwork with Walcrete and later plaster it to match in. Again, it might be more convenient to corbel a broader base to seat the beam ends, especially if the piers or walls are of good stone or old brick. To do this saw two pieces of board of the same length and tack them across the corners of the beam and vertical post, at the same angle. Having made sure the vertical side of the angle is well keyed for bonding, build the corbels at either end with stone and Walcrete, employing the boards as forms. Trowel to a tight fit all round, then render the two corbel faces at each end. When the Walcrete has set remove the boards and plaster to a smooth finish.

Now that the exposed beam or joist is secure, remove the nails from it, give it a good sanding, and treat it with at least two coats of insecticide stain. Make good the ceiling tight to the beam with 'Thistle' or other gypsum plaster and Walcrete any ragged edges, where the wall has been removed, to within half an inch of the wall surface. When it is set, trowel flush with plaster.

If we want to remove the partition wall and the beam is unsafe to carry the weight above, start by measuring the length of the beam, and prepare a replacement. As with the other replacement timbers, it is quite possible to find something suitable round a farm or, failing that, from a rural woodyard or local demolition contractor. With your 'new' beam ready and well soaked with insecticide, the next step is to shore up the ceiling on either side of the old beam with two long stout planks, one in each room and supported from the floor by lengths of $3 \times 3$ in timbers. Cut the timbers a foot or so longer than the vertical height so they can be tapped snug, thus taking the weight while angled a little outwards at the bottom—then you can get the replacement directly underneath. Now take away the wall partition carefully, so as not to disturb the surrounding structure more than is absolutely necessary.

Cut out the condemned timber, or release it from its seating at one end and carefully work it free, and set the replacement in position. This particular process will probably need some strong muscular assistance. Lever or prop the beam up tight to the ceiling and bed the ends well in with Walcrete, corbelling if necessary to make sure the beam offers firm support all along its length. When the Walcrete has hardened firmly you can ease away the shoring and make good, as shown before.

## ENLARGING OR KNOCKING THROUGH A WINDOW

Old houses, and particularly old cottages, were seldom if ever designed with large windows. Perhaps it was because of the price of glass, or because ordinary casement windows were made to a more or less traditional style by the local carpenter, who worked on the principle that the smaller the frame and opening the easier it was to secure it against the weather. Even though sliding and upper sash windows were in quite common use in many houses from the early part of the eighteenth century, the more rural dwellings preferred the small wooden-framed casement with one half hinged to open out. In many stone areas it was cheaper to construct a simple fixed stone mullioned window than to fashion one of wood with mortised glazing bars and hinges. The leaded lights were so fitted to the mullions as to be removable if the owner moved to another house. Original iron casement windows are still to be seen in some old cottages, particularly in Derbyshire, where many date from the seventeenth century.

But we will here accept that the windows we are interested in are the common wooden casements. Perhaps one particular window in the lounge would, if it could be enlarged, not only throw more light inside the room but widen the view of the garden. Let us see how it can be done.

It is best when restoring the outside of an old cottage to retain as much of the original character as possible. Improve it by all means, but try not to make any alterations to the general structure that will conflict with its intrinsic quality and old-world charm. Only too often one sees recent renovations which have, for the lack of a little extra thought, incorporated into old period cottages such glaringly modern additions as badly proportioned steel-framed windows, pointed-brick extensions, multi-coloured asbestos roof tiles, and other up-to-date knick-knackery straight from Ideal Homes.

There is no reason why you should not enlarge your window space so long as you stay in harmony with the overall architectural line. Let us say, for example, that your present window frame is of wood, 3 ft 8 in high by 2 ft 8 in wide, and six-paned with one half opening.

Now you can double the size of the window by widening the area to take another frame of the same dimensions, fitted almost jamb to jamb. By extending the window horizontally it will remain in character with the other windows. But if you heighten the window, not only could you mar the general appearance by throwing the lintels out of line, but you could well run into difficulties when excavating above the existing lintel, if its condition is at all impaired. So the horizontal extension is to be recommended.

Mark the outline of the area to be excavated on both sides of the wall, allowing 6 in or so for a central pier between the windows on which to seat the lintel ends. Make sure the area is free from any vertical crucks or studding. Next prepare your new lintel and duplicate window; your carpenter or any joinery shop will make you up an exact duplicate of the present window. As to the lintel its width will depend, of course, on the thickness of the wall. Try to match it to the other. It may be that a couple of lengths of 3-in oak boards will suffice, again depending on the thickness of the wall for their width.

The wall will probably be of cob or stone. Cut out a slot to take the new lintel, tight up to the end of the existing one, and extend it far enough at the other end to allow for the central pier and tenon on the new frame. Cob is usually solid enough to stay firm while this operation is in progress, but with piled or mortared stones the situation can be more tricky. Sometimes, in order to cut a lintel slot to size, it is necessary to make a ragged cavity where a large stone, annoyingly seated in the middle of the line to be cut, has to be removed. Where this happens take a little extra care, prying the stone away gently so as not to disturb the surrounding area above. With brick you should experience less trouble. The lintel over the existing window may possibly extend beyond the 6-in pier intended as a central support. If so it will have to be cut back to a point midway over the pier to allow at least 3 in for the new lintel end to seat in. As soon as you have made sufficient room, set the new lintel in position and shore it up firmly all round.

Now cut down the vertical side at the extreme end where the jamb of the new window will fit and take away an area of wall inside the line to the lower sill level about a foot wide. Make the vertical edge good immediately by Walcreting the stones up tight under the lintel and trowelling to a smooth surface. Let the Walcrete harden before excavating further. Once the far end of the new lintel is secure cut down along the other vertical side so that you leave 6 in of the wall between the two window jambs as a central pier. Trim the edges and make them good with Walcrete, trowelling to a smooth finish and scoring a key for the plaster later. Now that the two sides will hold the lintel secure make good the top with Walcrete, pressing it well in between the top of the lintel and the broken wall edge. You can now knock away the remainder of the wall in the new window area and remove the shoring.

There is now only the lower sill to be made good before fitting the new window in position. Drive three or four anchoring nails through the jambs on each side and into the still green Walcrete, first checking that the window is level and plumb. Trowel Walcrete tight all round, allowing for a finish coat of plaster.

*Page 139* (*above*) Interior of restored sixteenth-century inn. Beamed ceiling and clavel have been exposed and fireplace excavated to original size, revealing inglenook. Present corbelled fireplace (built of six-inch rough-faced bricks) harmonises with its surroundings; (*below*) corbelled brick fireplace built into original area. Small semi-circular room on right was originally smoke-room for curing bacon

*Page 140* (*above*) Loft interior of converted mill; (*below*) exposed wattle-work, with daub removed, forming effective screen linking two rooms to make a spacious lounge. The stiles (or upright staves) are of elm, and the lacing is of hazel twigs. The lower sections of horizontal twigs were removed because of extreme deterioration. A small portion of the original daub was left *in situ* (top left-hand corner) and glassed in. The joist and board-beamed ceiling was revealed during restoration after taking out a false eighteenth-century plaster ceiling which had been built beneath it

## BUILDING AN EXTERIOR BUTTRESS

Sometimes a wall of an old house or cottage has become a little infirm, either through subsidence beneath or because of uneven settling of its composition; it may show signs of bulging or perhaps leaning out. In most cases this need not be taken too seriously, but if it is a matter of reassurance, then a buttress can be built to strengthen the wall against any further tendency to move out of line.

First dig out an area 5 ft square against the middle of the wall to be supported, and to a depth of a foot or so. If the soil is very loamy or spongy take it down a little deeper. Fill the hole with stones and rubble to within 4 in of the top and bring it up level with fairly sloppy concrete (three gravel, two sharp sand, one cement). Poke the concrete mix well down into the hardcore, water it if necessary, to make a solid foundation.

Lay a lining course of old bricks or stones about 3 ft square tight to the wall on top of the concrete footing. Now run two lengths of parallel string to the corners and fix them to the wall directly above where the buttress will end, stretching them tight. Build up the sides of the buttress a few feet, following the string guides, fill with rubble, and pack with sloppy concrete. Repeat the process until, as the buttress tapers off, you find you can build with stones and concrete without the rubble filling. Render the buttress on three sides with cement, simulating as closely as possible the texture of the wall it adjoins.

Don't be too concerned about well-defined angles to

the edges of the buttress as long as the vertical is plumb. This is one case where, by virtue of your amateur status, you will in all probability make a more convincing and 'in-character-with-the-cottage' job than would a master mason. But it is advisable, before starting any work of this nature, to consult your building inspector for any comments he may have to offer.

## CENTRAL HEATING

A very important item to consider for installation in your cottage is central heating. It is odd that so many people regard central heating as a recent innovation. Many Americans are, in fact, convinced it was their original idea. One Californian housing expert told us with much pride that, not long before the last war, he was partly responsible for introducing a brand-new system of central heating. He called it 'Radiant Heating': warm air, heated by an oil-fired burner in the basement, was inducted through a series of thin tubes which ran under the floors and inside the walls, thus ensuring an even temperature in every room. It was, he maintained, the last word in housing development. Apparently he was unaware of the fact that the Romans used exactly the same principle, known as 'hypocaust heating', in their buildings 2,000 years ago. The only difference was that they relied on charcoal furnaces, situated away from the main building and at a lower level, to heat the air which rose and circulated throughout the hollow walls and floors by means of concealed earthenware pipes, or 'cuniculi'. Evidences of hypocaust heating can be seen today in excavated

Roman villas about the English countryside. Though an attempt to revive the old Roman system of central heating was made in 1908 by Professor A. H. Barker, who patented several variations of the principle in this country, it was another thirty years before the idea was to be considered generally acceptable and worthy of wide-scale promotion. So rather than being something entirely new, central heating is really an old idea just borrowed back.

Of the many systems of heat supply available today, hot air, oil, gas and electricity, we recommend, providing the service is laid on, electric storage heaters. Of course, any of the others would probably do as well, but the great advantage of the electric heaters is that little or no structural disturbance is involved, and consequently no mess. It is a relatively elementary matter for the fitters to run their wires around the skirting, or to drill a finger-sized hole through a wall. Furthermore, off-peak storage heaters can be fitted after all the decorating is done so that their particular size, colour and positioning can be determined at leisure to conform with the overall house-plan.

And while on the subject of fixed heaters, remember that walls in their immediate proximity, which have in all likelihood never been subjected to constantly warm, dry temperatures, may possibly shrink a little to begin with on this account. If cracks should occur simply fill them with a cellulose paste and touch up afterwards.

If you do decide to have a radiator heating system installed, choose the slender-type radiators and paint them to match the wall behind so they will not too much affect the character of their surroundings. Also try to avoid having trails of pipes running round the walls at awkward heights, and see that any piping which has necessarily to pass through walls does so as unnoticeably and with as little disruption to woodwork or masonry as possible.

sion paints. So before you decorate, go over the ceiling with a broadbladed scraper and remove as much as will lift easily, being careful not to gouge into the plaster. Then take a bucket of warm water and a soft brush and wash off as much of the remaining whitewash as possible. Fill any cracks with a cellulose filler.

Unless the plaster is particularly firm it is advisable to paper it with a good stout lining paper before you apply ceiling emulsion. First give the ceiling a coat of size to seal it and to make it easier for the paper to stick (a thinned proprietary brand cellulose paste is ideal for this). Paste the lining paper on in short handy lengths, and roll or brush each length firmly from the centre outwards to get rid of air pockets. When the paper is dry give it a couple, perhaps three, coats of white ceiling emulsion.

If you intend to paper walls which have not been previously papered, they will need sizing first. But if you are simply going to apply an emulsion paint, fill any cracks and nailholes with cellulose filler and sand smooth before you begin. On the inside of exterior walls, especially close to the floor level, where signs of damp have penetrated, even though the cause has now been arrested, treat the area with a proofing liquid before applying emulsion paint or wallpaper. But before using such a sealer remember to give any existing moisture a chance to dry out completely or you will only seal the damp into the wall, thus defeating the whole purpose of the operation.

It is well to bear in mind when considering your scheme for overall colour, decoration, and furnishing, that it must suit the character of the house; you are not dealing with a newly built house which as yet has no character—just smooth bare walls and new flooring.

In your old cottage the beams, the exposed woodwork or stonework, tiled floors, any panelling, etc, are leading characters and any decorative improvements should be with a view to displaying them to the best advantage rather than providing competition. So however much you lean towards a riot of colour in your surroundings, it is wisest—and this is from long experience—to keep all plastered walls, especially in the downstairs rooms, to one light shade of emulsion. We found, after trying almost the whole range of light colours from stark white, that magnolia is the shade that blends best with old woodwork. White tends to throw grey shadows in dark corners, creams become dirty-looking after one season, and various tones of yellows, pinks, and blues look too 'arty'. Magnolia is very light but has just enough pink in it for warmth, and in total effect is sufficiently neutral to blend with any other colour.

Old properties do tend to need the odd touching up from time to time, so that another advantage of using only one shade of wall colouring is that you can avoid any patchwork effect. One light shade, moreover, helps to create a sense of space and airiness in rooms that are fairly low-ceilinged and perhaps heavily timbered. Bedrooms are a different matter. They are more intimate areas, furnished for individual tastes. Cosiness is important and wallpaper comes into its own, preferably old-fashioned patterns in soft colours.

The kitchen and bathroom are, of course, going to be your two really modern rooms, and any labour-saving materials on the market will look quite suitable. One word here—bear in mind that tiling and uneven walls do not go together. Either see that the walls around the basin and bathtub, or the kitchen unit, are squared up, or stick to a good heavy plastic wallpaper.

If your inside windows look as though they might be very old, perhaps even original, with stout chamfered frames and mullions, wooden-pegged, covered by many layers of paint, try exposing a small patch. The chances are that the wood is good old oak. And if you have the patience of Job it is certainly worth your while carefully to expose all the wood with a reliable paint remover. Having restored the woodwork to its original appearance, wash it well with turps. When it is thoroughly dry rub in linseed oil and later give it a good polishing. An old inn we restored had two such windows. It took a lot of hard work to remove the many coats of paint, but underneath was the original oak surface, almost black with smoke. A vigorous buffing and polishing gave a satin-smooth gleam that amply repaid the effort.

Stonework inside the house is apt to look a bit dry and dull, particularly big stones around doorways and those in the piers supporting the clavel beam over some fireplaces. Scrub the stonework thoroughly clean with mild soap and water. When it is dry generously brush in colourless wax polish, taking care not to leave any surplus wax on the stone. A good brushing with a soft brush will bring up a glow. Finally, apply a light coat of a liquid vinyl polish—the colourless type used for floors—and buff it well as soon as it dries. This treatment helps to seal the more porous stones and gives a warm smooth finish to the whole.

Similar treatment works wonders with old tiled floors. There is nothing so enchanting as old quarry tiles, worn into hollows and dips, their surfaces smooth as silk in every shade of red from deep rust to pinky yellow. But without care they can develop a patchy, dull appearance. Here again a really generous application of colourless wax, well rubbed in followed by a light coat of liquid vinyl polish will bring up their colour and texture.

Your beams have already been treated and stained, but they will need plenty of wax polish, too. Leave them to dry thoroughly, then buff with a soft duster.

You will find that decorating and putting the final touches to your restored cottage is more satisfying than you could have imagined. For it is only when you reach this stage that you really begin to appreciate what all your hard work has achieved.

CHAPTER 11

# The Garden

---

One of the delights of owning a country cottage is to own with it the mature and often odd-shaped patch of land on which it stands. The garden of an untenanted old cottage somehow seems conscious of being part of the structure itself and, as if sensing withdrawal of interest, collapses in sudden decline, succumbing to the take-over bid from the long-frustrated weeds. Nothing looks so forlorn as a neglected cottage garden. By the same token, nothing looks so restful and inviting as the same garden carefully tended, with splashes of colour and winding paths leading to nowhere in particular.

There are so many interesting and satisfying things to do with a small country garden. The soil, for so long accustomed to being nurtured and conditioned, will readily respond to the slightest effort, unlike that of so many newer gardens whose ancestry is somewhat dubious. The old cottagers kept their garden simple. Most of the space was laid out in potatoes, cabbages and other vegetables, for economic reasons, with here and there clumps of homely flowers, herb-fringed borders and a few fruit bushes. It had a settled atmosphere of quiet neatness. Non-utility details such as hard-surfaced paths, orderly fencing or enticing retreats were seldom bothered about, as the cottager had neither leisure nor money to spend on luxuries.

But there is no reason why you should not add to the charm of the garden by putting into it your own touches, so long as you keep in mind its original character.

## PATHS

There are several types of garden path which are simple, and yet not too expensive or time-consuming to lay. Crazy-paving makes an attractive path and affords a firm, dry footing. Measure approximately the number of square yards of material you will require and order it from your local builder's yard or nearest quarry merchant, to be delivered in broken paving slabs or flat stones. Level the path, which, if an old one, is probably rough-cindered, allowing for the 2 in thickness of the paving plus 1 in of sand base. Set the pieces of broken paving in the sand, arranging them in interesting designs. Check with your level that they are firm and even, and spaced so that there is ½ in or so of clearance all round. Repeat this procedure over the whole area to be paved. You can now either fill the gaps with a fairly sloppy cement or pack them with soil, to be later sprinkled with grass seed. If you use cement fill the gaps to within $\frac{1}{4}$ in of the top so that the overall staggered pattern will show off to better effect. Wipe off any excess cement from the edges while it is still wet.

You may prefer to make your own crazy-paving slabs. To do this, first make several different-sized 'forms' or frames, by nailing together, in the required shape, lengths of 2 in batten, setting the forms either on level soil or on top of a few thicknesses of old newspaper. Mix up a batch of concrete (three parts small gravel or chippings, two parts sand, one part cement). Line the inside of the forms with greaseproof paper, or coat them with grease, to prevent the concrete from sticking, then fill with concrete and trowel level. When the concrete has set simply tap away the slabs from the forms and refill as many times as is necessary.

Old bricks, arranged in patterns and bedded in sand, also make a solid and interesting garden path.

If you happen to be in a district where smooth pebbles are plentiful, you could make a cobbled path. Again see that your foundations are firm and level, and have at hand a bucket of soil to pack up any pebbles which may need raising so as to bring their tops to an even height. Set the pebbles snugly together, on their thin edges, and proceed along the path. An assortment of coloured pebbles can be made even more interesting if arranged in regular patterns as a mosaic. When you have finished, sprinkle a layer of fine soil over the cobbles and then sweep it off. This allows some of the soil to work down between the pebbles, acting as a binder.

When making new paths it is well to remember that there is no art in a straight line. Set them out in attractive curves, sympathetic to the character of the garden.

## RUSTIC WORK

There is usually to be found in an old cottage garden a small, crudely fashioned log-shed, often so dilapidated as to be beyond repair. Since an outside shed is invaluable for storing garden tools and other odds and ends, rather than do away with it altogether you could replace it with a rustic shed, something you can easily make yourself.

If there are not sufficient poles for the purpose to be cut from trees round about, or perhaps from the prunings of a neighbouring orchard, the local coal and wood dealer can probably arrange to have a quantity delivered. The majority of lengths should be reasonably straight, about 2–3 in thick and roughly 8 ft long, with a dozen or so a bit stouter. The remainder can be made up of varying lengths of forked and gnarled pieces. Almost any wood, apple, pear, larch, fir or sweet chestnut will suit admirably. Beech is not to be recommended as a first choice because of its susceptibility to quick rot. Willow and poplar have a tendency, if untreated and staked into damp ground while they are fresh cut, to take root and start new growth. Incidentally, this is a point worth knowing, perhaps to advantage, for rustic lattice-work. It is always better to nail rustic-work while it is green; dead wood is difficult to nail and splits easily. The only tools you will need for the construction of the shed are a hammer, coarse frame-saw, measuring tape, sharp hand-axe, spade, creosote and plenty of mixed 2–4-in nails.

Assuming there is already a floor to the shed, possibly of brick (otherwise you can put in a raised board floor later), first stake into the ground, after treating the bottoms with creosote, four stout uprights, clean cut at the top, tight to each corner. If the roof is to be a simple lean-to, arrange the uprights accordingly with the back higher than the front. If, on the other hand, you want a gabled roof, the four corner poles will need to be the same height. Now drive three more uprights into the ground, with the buried parts also treated with creosote, between and in line with the others, leaving the door side to be staked with two poles spaced for the doorway. Nail four lengths to the tops of the uprights all round, allowing for the two gable-end poles to be extended 1 ft beyond the tops of the corner uprights, running back and front, for roof clearance, having notched the overlapping joints beforehand, to complete the frame. Saw off any excess. Now build up the walls by nailing straight poles, horizontally, on top of each other all round, interlocked and notched at the corners to ensure the minimum of gaps. Allowances should be made of course for the door and any windows deemed necessary.

In the case of a lean-to roof, allow the overhang on the low side only. Fit a ridge-pole across the tops of the taller uprights and nail on to it several rafter lengths, parallel and equally spaced, down the roof angle. Secure the other ends to the lower horizontal of the frame. Starting from the bottom, nail the thinnest and most evenly proportioned poles snugly together all the way up to the ridge. Repeat this procedure up the faces of the half gables at each end. Trim off the surplus ends.

For a gabled roof first build two trusses by measuring and fitting two couples of matched poles, crossed and notched close to the top. Nail the bottoms firmly to the corner posts at each end and fit a ridge-pole across so that it sets firmly over the crotches, then saw off all excess lengths. Fit the rafters on each side, as with the lean-to roof, and build up the roof face and gables with thin poles. Short lengths of misshaped or gnarled pieces can be added for ornamentation. Line the door and window openings with stops, and either make your own door and frames for glazing, or buy them.

To waterproof the roof, tack on horizontally laid roofing-felt, secured by strips of thin batten. Start from the bottom and allow a couple of inches of overlap, seeing that the lower edge is tucked under the edge immediately above it. Any further dressing up, such as applying bitumastic paint to the roof or lining the inside of the shed, is a matter of choice.

It may be felt necessary to proof the rustic-work against worm or rot. If so it is advisable first to remove all the bark so that the creosote can soak well into the wood. You must decide, though, whether removing the bark so impairs the rustic appearance that it is not worth the effort.

A summerhouse can of course be put together in the same manner as a simple shed. Rustic fencing or trellis, ornamented with climbing plants, not only makes practical screens, but also enhances the charm of a cottage garden.

## BUILDING A GARAGE

It is highly unlikely that an old cottage will have a garage, as such. And a garage is a necessity, whether or not you build it yourself. Choose your site carefully beforehand so that its position and design will blend in as harmoniously as possible, in keeping with the lines of the general overall architecture. A garage is really an elementary structure, three walls, a roof, floor, window and ready-to-fit door, so that its construction is not a vast undertaking. Of course your local council will naturally want to approve the site and rough plan before you commence, but this should not, under ordinary circumstances, prove too difficult or cause much delay.

Whether you work with brick or concrete blocks the principle of wall building is much the same. Draw a plan of your proposed garage and order the materials accordingly; your builder's merchant will be able to help you a great deal. Prepare the foundations by marking out and digging a trench for the footings. Providing the soil is reasonably firm, a trench 1 ft deep and the same width should suffice. Now fill the trench all round with a 6 in base of concrete (four parts small gravel or chippings, two parts sand, one part cement). When it has hardened you can start the brick or 4 in concrete-block work, using a mortar of four parts fine sand, one part cement. Start with the four corners, using a builder's square to set your right angles, and build up to floor height. Lay on a trowel of mortar and smooth it more or less evenly to about ½ in thickness. Mortar the bonding side of the brick or block and set it in position, tapping it firm and square.

The simplest method of laying bricks is stretcher bond, ie bricks laid on their flat lengths with the 'frog' or hollowed part uppermost, and each joint breaking below and above the middle of the brick between. To accomplish this it will be necessary to cut every other brick in half, 'half-bats', alternating as you build up each course. Start your pattern right and you will experience little trouble in keeping to it, so long as your lines are true.

The most satisfactory brick wall is an 11 in cavity wall, that is, a double wall with a 2 in clearance in between, reinforced with wall-ties built-in from the outer to the inner wall and spaced about 3 ft apart horizontally, and 1 ft 6 in vertically. Laying 18 in × 9 in × 4 in concrete blocks is a less complicated method of walling and shows quicker results than brickwork. Also it is every bit as satisfactory, especially if you intend to render and coat the exterior with a cement-based paint. So, while it is perhaps helpful to have a basic understanding of how to lay bricks, it is well to bear in mind that (a) the chances are that a new red-brick structure in the proximity of an old-walled cottage will look jarringly out of context, and (b) a structure built of concrete blocks, cement rendered and painted in to match the main building will probably prove more satisfactory all round. Therefore we will, in this instance, stay with the larger proportioned blocks.

With the four corners laid up to floor height, lay on strips of damp-course and continue to build up your corners, or quoins, to a height of 3 ft. Fix a tight string-line from each corner all round so that it is about ¼ in clear of the outside face, checking with your level to make sure the lines are true. Now build up all round the footings so as to come even with the tops of the quoins, remembering, of course, to lay the damp-course in position as you proceed. Constantly check with your level. After laying each block scrape off surplus mortar and point the joints by drawing the point of your trowel along them.

Let us assume the dimensions of the garage are to be 18 ft × 9 ft, a wall height of 7 ft 6 in and 4 ft to the ridge. Allow for a window of, say, 4 ft × 3 ft and a lift-up door 7 ft × 6 ft 6 in. You will require approximately 350 concrete blocks for the construction work.

At a point midway along each side wall bring up an extra block as you build, cemented on the inside, to make a pier for a central roof truss. Fit your window frame as you proceed, setting either a concrete or wooden lintel immediately above it. Be sure to allow the exact measurement for the garage door, after lining the sides with 3 in × 4 in timbers well secured to the edges of the blocks, and a stout 6 × 3 in lintel over the doorway. When you reach the wall height lay a length of 4 in × 2 in wallplate on top of each side and secure it firmly with heavy nails into the wet cement. Now build up the gable ends, being careful to see that all the angles are equal. As you proceed set two 4 in × 2 in purlins the length of the roof and equally spaced on each side, the lower one a little less than half-way up and the upper within 1 ft of the ridge height, rebating the ends into the gable wall so that they set firmly and flush with the roof slope.

To give the roof rigidity make a jointed truss of 4 × 2 in timber in the shape of a triangle and fasten it on top of the central piers so that it spans across to the wallplates on either side, thus giving support to the purlins. Lengths of asbestos or aluminium sheeting can now be laid on the roof, overlapped at the edges, and drilled for nailing through to the purlins. Roof sheeting can be obtained with various surface finishes, so choose that which seems most suitable to the character of the surroundings.

Guttering boards, guttering and downpipes are

comparatively simple to fit. Nail the guttering boards into the cement joints while they are still green and trowel a little extra cement around them to ensure a firm hold. Remember to angle the guttering slightly, to allow water to run towards the downpipe corner.

It is advisable to render the outside walls by plastering them with a thin coat of cement, using a flat trowel, in order both to proof them against damp and to give them a more mellowed appearance when Snowcemmed. Don't worry about trying to achieve too even a surface. Frankly, the more unprofessional it looks the closer it will be to assuming a naturally rural effect. As to the garage floor, this can be put in last of all so that you are able to work protected from the weather. Dig out the floor to a depth of 4 in, fill with concrete, allowing for a slight run towards the front end, and trowel it smooth.

## GARDEN STONEWORK

The charm of an old-fashioned garden can often be enhanced by incorporating a little unostentatious stonework, such as low cleft-walling along path edges, up the sides of stone steps or around old, and perhaps dangerous, open wells. There are various types of walling stone to be had, two of the most effective, apart from simple garden stones or pebbles, being laminated blue lias or yellow-brown Purbeck slabs which, when broken into rough sizes averaging $1\frac{1}{2}$ in thick and 5 in across, are easy to handle.

First lay a couple of inches of concrete footing and then cement each piece of stone lengthwise, rough-faced on each side of the wall. The wall thickness should be about 9 in. You will find that the stones make their own interestingly irregular bond as you build up. The joints can either be deep pointed or left flush. For a low wall leave a hollow at the top about 6 in deep and later fill it with soil. Rock plants will grow readily in the cleft to spill over the edges in a most attractive manner.

Slender stone pillars for garden arches or climbing roses are another feature becoming to a small garden, and are simple to build. All that is required is a length of straight old iron pipe to use for a core. Drive the pipe into the ground and build up tight around it with small stones, well cemented together. Before the cement is thoroughly dry brush off any excess cement with a stiff wire brush.

Dry-stone wallking is an ancient craft which has all but died out in most parts of the country, though there are a few districts, especially in the Oxfordshire area, where it is still practised. Flat stones are simply stacked one on top of the other to the required height, but there is of course more to the process than meets the eye. The secret of building a wall of rough stones without any bonding whatsoever, neither mortar nor mud, which is yet so substantial as to stand, impervious to the elements, for centuries in many cases, lay with the almost instinctive sense of balance and interlocking principles the old craftsmen possessed.

An effective and sturdy modification of true dry-stone walling can, however, be constructed by the amateur. It is important, of course, to use flat and fairly even-sized stones. Each stone must be settled in position all round, the layers interlocked and bedded on a pad of moist soil. Any gaps should also be packed with soil and tamped firmly. A good dry-stone wall needs to be at least 14 in thick. To face a bank with dry-stone walling, a more satisfactory result is to be obtained by first angling the bank a few degrees off centre so that it leans back at the top. This way you can build up the wall a single stone thick, bedding each layer as you proceed and packing plenty of soil behind the stones. After a little time they will settle in and the wall will remain solid. To give it that little extra touch, plant seedlings of colourful alyssum and aubrietia in the soil between the stones. Hardy wallflower seeds will also flourish. One method of planting seeds in the cracks of an earth-mortared wall is to blow them in with a pea-shooter, poking a little soil in on top of them.

CHAPTER 12

# Conclusion

A thoughtfully restored old cottage can be said to have one great advantage over the up-to-date technically designed and scientifically constructed house: it has already been a home, in the truest sense of the word, for many families over many generations. And over that long period of time it has become saturated with a warm, lived-in atmosphere which gives it a personality all its own. Is there any wonder? Every item that went into the original building was first sought for, shaped and prepared by hand and then fitted into position, more by painstaking trial and error than by any pre-calculated precision. Time meant little to the early builder. He kept no accurately balanced books to show expenses and labour charges against his expected profit margin; he had no office overheads or disputes over wages to contend with; no deadlines to meet. It was not for him simply to pick up the telephone to order, almost in a sentence, the exact kinds and amounts of mass-produced materials necessary for the complete construction of 'House-plan 2a' or 'House-plan 4c, with or without garage', then to pass on various phases of the work to outside subcontractors. The old-time builder felt his way along, bit by bit, absorbed in his work. If a timber seemed more pleasing to his eye to be angled this way instead of that way, then, providing its function was satisfactory, even though it might be out of line, that was justification enough for him to carry out his whim. His technique was primitive, his execution spontaneous. The result was a form of unconscious art.

There is no time now for the old methods of handicraft. A machine will produce a reasonable facsimile in a tenth the time at a tenth the relative cost. Today's demands must be met—and on time. Where once the painter would perhaps spend days carefully gathering and refining the basic materials for his colours, mineral nodules and plants from the countryside about him, to be compounded with walnut oil in a stone trough, his modern counterpart buys his wholesale-manufactured synthetic paints by the simple expedient of quoting the code number of the shade he requires. He has no need to know its composition. His sole concern is to apply the material furnished for its ultimate effect, as per instructions, and against the clock.

But while extolling the virtues and craftsmanship of the old-time builders of country cottages, at the expense of the modern house contractor, none of us would willingly go without some of the modern fittings and services, which have indeed become an integral part of our lives. Electricity, for instance, is now a taken-for-granted essential. Plumbing is another case in point. It is not so very long ago that, after the introduction of mains water to ordinary houses, supplies of untested water were carried by pipes of drawn lead. Apart from in a few districts where lead was to be found, and in the homes of the wealthy, water mains, prior to the growth of urban populations during the nineteenth century, were usually made of wood. Even pipes of pumps and wells were made from bored branches or tree trunks—durable alkathene pipes are a big step forward in a comparatively short space of time!

The amount of window space in smaller houses has grown steadily since the turn of the present century. Window glass used in England before the repeal of the excise duty imposed on it—which greatly hindered its wider use—was commonly known as 'crown glass'. If examined along its face it can be seen to have a wavy surface and is greenish in colour. The molten glass was blown into the shape of a flattened sphere and then spun into a flat plate called a 'table', of which twenty-four made up a 'crate'. The lump left in the sheet where the glass had been attached to the pipe was known as a 'bull's-eye' or 'bottle-end'. Bull's-eye panes are still to be seen in the windows of some old cottages. As C. F. Innocent points out (*Development of English Building Construction*), 'Although the bull's-eyes or bottle-ends were the cheapest and most inferior part of the sheet, they became fashionable in the so-called Queene Anne revival, and were bought from old cottages for re-use in good modern houses.' The cheaper mass-produced sheet glass is of course technically if not aesthetically superior.

The whole approach to cottage restoration must be

governed by a wish to retain all of the original structure that is pleasing, practical and of the true essence of a bygone period, while also incorporating the essentials to comfortable modern living. You will then have —literally—the best of both worlds.

APPENDIX

# References and Acknowledgments

---

Addy, S. O. *Evolution of the English House*
Aubrey, John, *Antiquarian Repertory*. 1678
—— *Builder, The*. 1847
Clayton, C. E. *Cottage Architecture*
Dawber, G. *Old Cottages and Farmhouses in the Cotswold District*
Innocent, C. F. *The Development of English Building Construction*
Moxon, Joseph. *Mechanick Exercises*. 1677
Price, W. F. *Homes of the Yeomen and Peasantry*
Shuffery, L. A. *The English Fireplace and its Accessories*

My thanks also for the helpful information I have received from the planning officers, librarians, archivists etc of the various county authorities I have approached, and to Mr A. J. Sparke of the 'Cottage Men' for his constructive criticisms and suggestions.

APPENDIX

# Government Improvement Grants Scheme

---

The situation regarding revised Improvement Grants is under review in the Housing Bill before Parliament at the time of writing. This Bill is to implement the policy for house and area improvement announced by the Minister of Housing & Local Government and the Secretary of State for Wales last April (1968) in the White Paper *Old Houses into New Homes* (HMSO), and embodies all the proposals in White Paper. If Parliament approves this Bill, the amount of the Standard and Discretionary Grants for improvements and conversions will be increased, among other provisions, and local authorities will receive new and more flexible powers.

At present, however, an Improvement Grant is not a loan. It is an outright grant of money from the local council, given in approved circumstances. Only in exceptional cases is there any question of repayment.

In order to qualify for a grant you must be the owner of the property, and the council must be satisfied that the house is likely to have a useful life of between fifteen and thirty years. The grant is paid when the work is completed to the council's satisfaction. If, after having received a grant, you want to re-sell, you are free to do so. But if the sale takes place within three years, for owner-occupation by anyone who is not a member of your family, the new owner will have to repay some of the grant, and this may well be a condition of sale.

To apply for a grant, simply write to your local council, advising them of your requirements. They will send you an application form.

There are two kinds of grants available—Standard and Discretionary.

A *Standard Grant* is towards the provision of certain basic amenities; local councils are obliged to pay it if the necessary conditions apply. These basic amenities are at present: (a) fixed bath or shower in a bathroom; (b) wash-hand basin; (c) hot and cold water supply at a fixed bath or shower, at a wash-hand basin, and at a sink; (d) inside WC; and (e) a satisfactory food store. If the Bill goes through (e) will be dropped from the list.

To qualify for a Standard Grant the house must have been built before 1945, or have been provided before 3 October 1961 by the conversion of a property built before 1945. Also (and this point is dropped in the Bill) it must, after restoration, have a life of at least fifteen years; any local council is likely to want this reassurance before approving a grant.

In certain circumstances it is possible to get more than the normal maximum Standard Grant, such as when it is necessary to build on to the house in order to have a bathroom, when a septic tank must be provided because no main drainage is available, and when a piped water supply has to be brought in.

How much? A Standard Grant is half the cost of the improvement works, including professional fees, subject to a normal maximum of £155 (the Bill aims to increase this to £200). The upper limit payable in the exceptional circumstances mentioned is to be increased from £350 to £450.

The second type of grant, the *Discretionary Grant*, is for a higher standard of improvement than that specified in the basic requirements for the Standard Grant. This is made entirely at the local council's discretion, and the 'musts' are stiffer but should present no problems if your restoration is properly carried out. After improvement the house must:

(a) be in a good state of repair and substantially free from damp,
(b) have each room properly lighted and ventilated,
(c) have an adequate supply of wholesome water laid on,
(d) have efficient and adequate hot-water supply for domestic use,
(e) have internal WC, if practicable, otherwise a readily accessible one outside,
(f) have a fixed bath or shower in a bathroom,
(g) be provided with a sink, or sinks, with suitable arrangements for the disposal of waste water,
(h) have proper drainage system,
(i) be provided in each room with adequate points for gas or electric lighting (where reasonably available),
(j) be provided with adequate facilities for heating,

(k) have satisfactory facilities for storing, preparing and cooking food,

(l) have proper provision for storing fuel (where required).

You cannot normally get a Discretionary Grant for work costing less than £100, but if you have already had a Standard Grant you can apply, within three years, for the extra amount if the two together come to at least £100.

*How much will you be entitled to?* Councils have discretion to pay up to one-half the estimated cost of the work (including professional fees), as approved by them, up to a maximum of £400. With the passing of the Bill this will be increased to £1,000. A word of warning. *Do not* start work, if you are applying for a grant, before the local council have approved your application, or you might be disqualified from receiving one. Wait until you have the notice in writing.

This is only a brief précis of what the Improvements Grants Scheme offers for the restoration of private property. There are wider provisions for flat conversion, premises for multiple occupation, etc. There are, moreover, possibilities of getting further financial help from local councils in the form of loans in particular circumstances, and for payments of a Discretionary Grant by instalments, instead of at completion, if the council consider this appropriate.

Such points are applicable to particular cases and have to be gone into individually.

Footnote:

The Housing bill referred to above has since been passed and came into operation 25 August 1969.

# Index